BLOOD SWEAT AND TEARS

FARZIN MOJTABAI

ISBN # 978-0-6151-7176-0

TABLE OF CONTENTS

Chapter 1

Activist Intuition

As the roar of the Tiger economies in Asia and abroad awaken the world to new global possibilities, Sweatshops remain unresolved and buried by business as usual. The champions of corporate led Globalization carry a philosophy that boils down to profit lines, rationalizing it as beneficence for impoverished work forces. Sweatshops are a largely known, but accepted tragedy in the Globalization era. Hope for the worst affected workers has at times, found momentum in a movement to stop them.

The human cost of labor has seen historic moments of influential advocating, whether the abolitionists of slavery or the formation of the International Labor Organization (ILO) in 1919.[i] Even slave resistance of antiquity was in essence a labor movement of a people that often reached revolutionary levels. The swords that led the charge into these movements were essentially activists harboring a fundamental intuition that human rights were being undermined. The job today is still too big for individual heroics of organizations and the solutions exist in a dynamic confluence of social, economic and political harmonization. This will be found in corporate accountability, the responsibility of governments, accessibility for consumers and the empowerment of workers.

The activists leading the charge at the dawn of the 21st Century make up a diverse cadre of anti-corporate globalization groups. Whether it's muckraker journalists, student activists, labor economy philosophers or humanitarian efforts, it's activism that shares a

gut instinct that there is injustice. What unites many factions of the movement is a

realization that Sweatshops are a viral symptom of negligent greed in economic

globalization. By confronting this Goliath their fight has just begun, but an achieved

awareness in the decade of the 1990s uncloaked the practices of powerful multinational

corporations sweating profits off a new form of slave labor.

The debate is still open as to when sweatshop labor can be considered a form of

slavery, but terms such as "bonded", "indentured" or "forced" clearly imply some level

of enslavement. This reflects a larger problem of semantic word play, interfering with

assuaging human rights violations on a number of levels. Finding an appropriate

definition of slavery is a continuing challenge for organizations and essential if proper

codes are to be agreed upon in legislation. Where it cannot be said that labor is slavery

when a wage is provided, literature from the ILO frequently resorts to phrases such as

"slave-like" labor.[ii] In many instances around the globe, labor is cruelly forced and wages

cleverly manipulated so that workers become indebted to employers in amounts more

than they earn. Most recently this has been seen on the Northern Mariana Island of

Saipan, where workers forced to pay recruiting fees became indebted beyond their wages.

Video footage of the worker camps present a clear display of harsh conditions that

penetrate the deceiving terminology tossed around to hide realities.

The air of this debate must be cleared so fresh perspectives can see past economic

agendas and into the core problem of human rights. Government and nongovernmental

(NGO) Labor advocates, as well as academics have at times found clarification and at

others cluttered the debate with accusations. The argument still stands with some truth that corporate interests in developing countries of the Global South (South America, Africa, South Asia) can lift people from poverty. The problem therein is development agencies, such as some from the United Nations, whose goals should coincide with multinational enterprise, rarely breath the same reality. Instead, the efforts of corporate interests rarely enhance humanitarian motives and a rift widens as multinational agendas become what professor Robin Broad calls "a common enemy: corporate-led Globalization."[iii]

This enemy is fueled by mechanistic forces in an economy dependent on the created indifference of the consumers that propel it. Fortunately this is not a reality the majority of humankind wants to perpetuate, creating the potential influence of ethical purchasing power. This undeniable feature of an increasingly aware world populace is largely untapped, hindered by confusion and misdirection. This in effect creates the mass indifference of an already selfish consumerism that is sweeping the globe. Up until recent the flow of this human impulse of ethical responsibility got filtered into charitable causes that, while necessary in immediate relief funds, gets lost into webs of systemic corruption. This is most apparent in the underdeveloped Global South of the African, South American, Middle Eastern and South Asian countries trapped in impoverished states. A lack of capital investments in infrastructure, both social and industrial, continues to point a finger towards the powers behind Globalization. The greatest advance in understanding the issue has perhaps come in that many are starting to realize that we are the ones with the power to shape globalization, before it shapes us.

The View From the Bottom

Individuals within corporations and the consumers who generate their profits are unfairly locked in an economic modus operandi with deeply flawed origins. This is not to say that solutions should be anti-capitalist or completely oppositional to multinational interests, but Corporate Globalization's frighteningly influential growth and power should be questioned. An increasing gap between rising middle class incomes and poverty rates in countries like China, India or Brazil reveal what critics have long seen as ruthless progress on the backs of laborers. All the while working towards building a better life, it inevitably turns out that they are building someone else's pyramid of wealth with little reward. The same pressures of a globally competitive marketplace driving this class chasm reflect the thinking that led U.S. and E.U. businesses into rationalizing sweatshops. Without appropriate regulation, initiatives and incentives to sway economies from this path, they'll get caught up in an economic jungle with thorny roots. Throughout the industrial revolutions of America and Europe, sweatshop conditions and child labor were seen as a byproduct of progress. While these countries may have achieved superpower positions, it was mostly by military might and a thorny past of cheap labor was their dependence, much more than their asset. Of course the economic knowledge of thinkers such as Milton Friedman file this class of labor as an asset of prosperity. In a truly idealistic concept of globalization, nations would exchange history's lessons for a more synergistic future.

Columnists and legislators toss the term "Race to the Bottom" into this scenario quite effectively, but more is often said than done. As the industrialized nations gear up for competition from developing countries in the race, prospects for better labor standards dwindle. The stakes for Labor and Environmental conditions are considerably higher now

with China's long awaited entry into the World Trade Organization (WTO) in 2001.

Though China itself has vast resources of labor where freedom of association (unions) is

non-existent, they are also positioned to tap the resources of a less developed Southeast

Asia. China is in a pivotal stance to either exacerbate the Global race to the bottom or

uplift the region's poverty. This is ultimately dependent on a re-imagining of WTO

policies to leverage their influence on member states in humanitarian efforts. Corporately

structured development in countries of need can create factory employment or

cooperative networks of suppliers that are at once profitable and humanitarian. Creating

the potential for this reality is the most immediate solution in preventing slave-like

sweatshops and rests on the imperative of consumer choice. Though it is more than their

purchasing power that will build this voice and their democratic vote is of larger

importance.

Viewing this solution from the bottom line for corporations is a risk too big to take

initiative on, nonetheless a few examples can be shown. Governments are even at the

mercy of economic interests and legislators haven't seen challenging unfair Free Trade

laws as a risk worth taking either. From a critical standpoint it may be more that

government players are incapacitated by a corrupt system from harnessing democracy as

a tool towards legislative solutions. Politics can't be completely written off, for example

in 2003 the state of Maine passed a statewide Anti-Sweatshop bill that bans all

sweatshop-produced goods from government procurement. It may be several years until

Maine's *State Purchasing Code of Conduct* realizes full impact and consequently other

U.S. states follow suit. Even if all 50 states enacted a Code of Conduct on government purchasing, the billion-dollar retail industry of the U.S. alone could perpetuate sweatshop labor. Just as legislation in America, the European Union or in any democracy must be constituent supported, corporate decisions must be supported by consumer reaction. The positive response of consumers will trigger a domino effect, as corporately led Global hegemony can instead become corporately structured development.

Signs of a disapproving marketplace are seen not just in the awareness of activism, but also in the consumer willingness to participate in proactive solutions. Take for instance a survey of 1,000 consumers done by Marymount University's Center for Ethical Concerns, which found that 86% of people asked would pay an extra $1 on a $20 garment if it was guaranteed to be made without sweatshop labor.[iv] Among the many other revealing responses of those surveyed, it's apparent that corporations are dealing with a concerned, partially aware majority that wants to see effective change in the world. Some consumers are even inspired enough to boycott certain labels, as seen in the past with Nike and Disney apparel brands. Activists may ruffle the feathers of sleeping giants like Nike, but to this day they lie comfortably on a bed of profits sweated off labor atrocities.

Born Again Labor Activism

Much of the public awareness in the 1990s was spurred by the likes of Jeff Ballinger who ran an office for the AFL-CIO labor organization in Jakarta. Ballinger's revelations of Nike paying Indonesian workers just a $1 a day led him to start the organization Press for Change, eventually influencing numerous other campaigns that continue to scrutinize Nike.[v] What does a $1 a day to a worker in Jakarta mean when there are little or no opportunities to live on? This is where a corporation like Nike can claim benevolence in bringing needed jobs, but in a different light it is exploitation of a disadvantaged people.

Even in a largely corporate controlled media, it's quite easy to bring shock value to such news and create a public lambasting of perpetrators. As news of labor violations accumulates exponentially, it leaves activists somewhat vulnerable to the question of relativism. If there were no jobs in a region to begin with, should the practices of corporations, regardless of standards, be tolerated? From a human rights standpoint the answer is a resounding no, but the problem of standards in global labor rights is an obstacle in creating a unified code of conduct for manufacturer compliance. As easily as activists can brew the news of atrocities in the global factory, economists can spoil alerts by arguing a region in need of jobs, leaving consumers troubled by debate discrepancies.

The anti-sweatshop movement is a main source of fodder in the mockery of corporate culture. The art form of activism also resurged during widespread student movements in the late 1990s, also spearheaded by an AFL-CIO summer internship program. Unbridled youth armed with the sagacity of union ethics established a frontline against rampant consumerism. The word "sweatshop" evokes a sinister atmosphere prompted in part by grassroots campaigns from students and dedicated human rights organizations that maneuvered their message through a corporatized mass media.

The antics of activists tainted corporate reputations and Anti-Globalization protestors tore into the headlines with intensity unseen since the Vietnam War Resistance. That is, at least in the United States. The media long overlooked how international labor and human rights upheavals might have been direct consequence of U.S. political or economic activity. Obviously the U.S. is not alone in their endeavors of global market positioning and the protests staged at World Trade Organization (WTO) forums targeted

the corporations of several powerful nations. The melting pot of movements that gathered

for the "The Battle in Seattle"[vi] protests on November 30th 1999 firmly pitted anti-

globalization activists against Free Trade policies in the public eye. There were so many

varying platforms of protesters that the focus of what was at times a riot became muddled

in the media and action networks. The sheer force of the protests, in Seattle and other

WTO assemblies, was equally chaotic, especially when attempting to define what it

means to be anti-globalization. Economist Joseph E. Stiglitz points out that the tools and

dynamics of globalization provide anti-globalization activists a resulting connectedness

that enables them to effectively protest.[vii] On the surface this seems hypocritical of the

anti-globalization movement, but is more ironic than anything and avoids the specifics of

such a criticism. While the Internet, cell phones and travel ability are all made accessible

through globalization, it is not the essence of what protestors attack. It is the policies of

world institutions often influenced by one-sided corporate interests. On the upside,

Professor Andrew Ross in his book *Low Pay, High Profile: The Global Push for Fair

Labor*[viii] presents a hopeful premise that the struggle for fair labor practices is a paragon

in the anti-globalization movement.

    To proclaim a stance against sweatshops is for an individual a position historically

rooted, morally inclined and culturally rich. By connecting with a spirit that has

transcended slavery, it carves a new socio-economic philosophy into our media saturated

zeitgeist. The ever complex mingling of language props up yet another semantic hurdle

for this philosophy in the use of "anti-globalization." In its broadest sense, globalization

is just the sharing of ideas across political and trade borders, whether in the tangible form of commodities or abstractly as theory. It is in theory that this concept has a very

appealing ring of an enlightened civilization existing harmoniously. Though when

theories of the political economy began to dictate the control and distribution of

resources, the criticism arose that not all aspects of globalization operate in the best

interest of all.

The flow of development money as navigated by the World Bank and International

Monetary Fund (IMF), forms an institutional trio with the WTO disharmonious to the

people who they proclaim to aid. On this tricycle of trade that globalization rides, it is

hard to say which institution is more of the squeaky wheel in need of oil. In the tides of

globalization critics stand firmly in the sinking sands of a changing world, gazing into an

approaching horizon and asking whose world is this? Finance Philosopher, George Soros

points out dilemmas such as that of WTO trade rules being disharmonious with ILO

conventions. For instance the WTO prevents export/import products that are essentially

the same from being considered different based on their production methods.[ix] With this a

rich country producing textiles could have advantages over a developing country

producing the same textile. This opens a global trade loophole that conflicts with ILO and

humanitarian efforts in trying to set labor standards for manufactures.

Before heavy organizing from NGOs and eventually legislators, prompted President

Clinton's administration to adopt the issue of Child Labor in the nineties, it was

immigrant communities that poured foundations of change. America's sweeping history

came upon the shoulders of immigrants and throughout their struggle for fair labor, they

rooted a thorn of solidarity in the side of corporations. An intimate account of especially

the women workers from author/activist Miriam Ching Yoon Louie reveals "…largely unsung heroines" that were "…risk takers" on the "…bleeding edge of anti-corporate movements in the age of Globalization."[x]

Spanning a historic struggle, Immigrant organizing was midwifery to the activism born of organizations like the National Labor Committee and Global Exchange. As Louie notes of these early labor movements, it was the groundwork for today's activists who "took the fight to GAP, Nike and Guess."[xi] This re-emergence of Sweatshops in public dialogue set afire an activism in a nation with an affinity for Labor struggles. As author Noam Chomsky has pointed out in interviews, the international celebration of worker's rights on May Day is largely ignored in the US today even though it is founded upon the struggle of US workers. This is probably due to its association of socialist ideals, a culture largely severed by the past 40 years in America. Today's generations have seen the full market transition to a Global economy and a movement that naturally turned an already critical eye to the Global supply chain. This was especially true for unions as jobs were being shipped overseas starting as early as the 1960s. Activists responding to factory outsourcing unearthed the infestation of human rights violations, presenting yet another layer to the global labor challenge. Meanwhile developing and under-developed nations striving to become industrialized or even just to create subsistent living conditions welcomed opportunities that foreign enterprises seemed to promise.

Looking back on this promise two very different stories can be told about what Global outsourcing does for the lives of workers. As for Domestic workers in the U.S. there are some critical of anti-globalization efforts conspiring that unions enflame the movement with ulterior agendas. Where workers in underdeveloped nations will work at bare minimum wages, unions fear this will lower the labor standards their predecessors worked so hard to elevate. This may be true, and rightly so, but distracts from the real issue for organizations more concerned with human rights violations.

As NGOs built support networks through existing unions or independent charities, horrendous reports of abuse and factory conditions poured in. Several reports reached congressional hearings in the U.S. congress and European parliaments, as concern turned outrage into action. Heavy campaigning in the media desensitized the victimization of workers, but to justify the appalling conditions was as criminal as the acts themselves. The Southeast Asian and Latin American regions became notorious for prison-like conditions at factories, as reports began to taint a label obsessed culture.

Clothing lines from Wal-Mart, JC Penny, Kohl's and Sears, along with apparel brands like Nike, Reebok, Guess, Polo Ralph Lauren, the Gap, Tommy Hilfiger, Sean John, Adidas and shopping malls filled more, become flash-in-the pan synonymities with sweatshops. That is, at least to those who sought awareness beyond their holiday shopping list. The outrage that inspired such impassioned activism wasn't just about living wages or fair trade, it was the egregious acts of slave-like treatment and violence neglected by corporations whose bottom line lived above international law.

Some of the worst reports arose from specs on the map engulfed by the vast Pacific Ocean, like the Island of Tafuna in American Samoa and the Island of Saipan. Both of which are U.S. territories that manage to avoid Labor laws, bringing executives looking for cost cutting into fertile land with factory owners. At the Daewoosa Samoa factory on Tafuna, manufacturing garments for JC Penny and Sears, Vietnamese workers were invited, charged employment fees up to $4000 (U.S), paid minimal wages and then subjected to inflated food and shelter expenses. On top of being enslaved in the factories premises, the reported abuses ordered by Korean factory owner Kil Soo Lee reached new

lows in November 2000. The workers had begun to voice their discontent and armed

guards were instructed to single out a young woman with a severe beating, leaving her

blind in one eye.[xii] Women workers at factories in Saipan faced similar conditions and

were coerced into signing contracts that resulted in forced abortions. With Several major

brands like Ralph Lauren and Tommy Hilfiger operating Chinese factories there,

lobbying blocked legislative efforts encouraged by Human rights workers. The lobbying

nonetheless of indicted corruption ringleader Jack Abramoff, who arranged pampering of

congressional leaders like Tom Delay on Saipan's beach resorts. Even after touring the

reality of the sweatshops, Delay spoke of the factories representing "everything that is

good about what we're trying to do in America, in leading the world in the free market

system", which appeared in an ABC News report.[xiii] If political leaders stay this far

removed from the issue when faced with the reality (Over 80 congressional reps visited

Saipan), consumers have a long journey to follow.

These stories emerged from distant, but official U.S. territories and Human rights

organizations battle even greater challenges in getting reports out of countries like China,

Indonesia, Honduras or Haiti. Activists of a new breed needed to create informant based

media to confront the secrecy of elusive multinational corporations. Human rights

advocates and student activists often resorted to physical protests in great strides of

empathetic grievances for a distant people. Corporations had to respond in compliance,

for when media opened the door to congress it allowed activists to engage a case for

workers. Their cases, when clearly documented, made denial of their claims and motives

near impossible. Yet wealth and power has omnipotence that can intercede with the democratic systems people rely on. In a decade of start-up organizing, hostile media wars

cracked the issue wide-open, as anti-corporate globalization became common knowledge.

The public conscience had been enlightened and the battle for their awareness and

empathy is still being waged today.

◎◎◎◎◎

[i] The ILO is one of the only lasting bodies originally started with the League of Nations, which dissolved in 1946. The United States never joined the League of Nations, but did become a member of the ILO under President Roosevelt in 1934. The ILO eventually joined the United Nations under a new charter, *The Declaration of Philadelphia* from 1944. www.ilo.org

[ii] In the 2005 *Report of The Director General* on *A Global Alliance Against Forced Labor,* Report I (B), the literature refers to "such terms as 'modern slavery', 'slavery-like practices' and 'forced labour' can be used rather loosely to refer to poor and insalubrious working conditions, including very low wages." pg. 11.

[iii] Robin Broad, "A Better Mouse Trap?" in *Can We Put An End To Sweatshops*, ed. Fung, O'Rourke and Sabel (Beacon Press, 2001), pg. 48

[iv] Michael A. Santoro, "Philosophy Applied 1: How Nongovernmental Organizations and Multi Enterprises Can Work Together to Protect Global Labor Rights" in *Rising Above Sweatshops: Innovative Approaches to Global Labor Challenges*, ed. Hartman, Arnold and Wokutch (Praeger Publishers, 2003) pg 106

[v] Liza Featherstone attributes Jeff Ballinger's findings as influential in the formation of the student activism group United Students Against Sweatshops. Liza Featherstone and United Students Against Sweatshops, *Students Against Sweatshops* (Verso Books, 2002) pg. 8

[vi] "The Battle in Seattle" or "...for Seattle" is the oft given title to the W.T.O protests in Seattle on November 30 1999. *The Nation, BBC News, CNN* and *Time Magazine* all used the title in press coverage and is the title of a book about the protests by Janet Thomas (Fulcrum Publishing, 2000)

[vii] Joseph Stiglitz, *Globalization and Its Discontents*, (W.W. Norton & Co., 2002) pg. 4

[viii] Andrew Ross, *Low Pay, High Profile: The Global Push for Fair Labor* (The New Press, 2004)

[ix] George Soros, *George Soros On Globalization* (Public Affairs, 2002) pg. 34

[x] Miriam Ching Yoon Louie, *Sweatshop Warriors* (South End Press, 2001) pg. 211

[xi] Miriam Ching Yoon Louie, *Sweatshop Warriors* (South End Press, 2001) pg. 228

[xii] *Report on the Working Conditions of Vietnamese Workers in American Samoa Report*, (The Vietnam Labor Watch in Washington D.C., Feb. 6 2001), http://www.vlw.org/#_ftnref5, (accessed 4/06) The case was introduced by the Vietnam Embassy in United States to the U.S. Department of State and Department of Labor in December 2000.

[xiii] "Forced Abortions & Sweatshops: A Look at Jack Abramoff's Ties to the South Pacific Island of Saipan and How Tom Delay Became An Advocate for Sweatshop Factory Owners", Narr. Amy Goodman, *Democracy Now*, New York, Wednesday, January 4th, 2006
http://www.democracynow.org/article.pl?sid=06/01/04/1524256&mode=thread&tid=25 , (accessed 01/06).

Chapter 2

Historic Regression

Human Labor is without question civilization's most valuable resource. One would be
hard pressed to argue otherwise, relying on absurdities that technological advancements
will free the labor masses. Perhaps in a Luddite's dystopia where entire economies are
automated, technocrats can render labor obsolete. Yet that vision is the product of angst
born of science fiction writers from industrialized nations. The reality of most the world
and especially the developing Global South is still an economy powered by laborers and
farmers. In parts of South Asia alone there are over 200,000,000 agricultural, industrial
and service laborers, not including the world's largest populations of India and Mainland
China.[xiv] Of these millions, there are close to 37 million working in what are known as
Free Trade Zones in Asia.[xv] The true bottom line for corporations is that people need
people for economies to operate, but still labor is viewed as expendable and the world as
just consumers.

As much as industrialization brings to mind leaps of technological progress, it's also a
historical regression in labor conditions. Even in so-called white-collar jobs, terms have
arisen like white-collar slaves, even white-collar sweatshops, reinforcing the notion of
perpetual corporate greed. Some argue that sweatshops of the industrial kind are a natural
stage of industrial development for an economy. In the competitive nature of market

forces, do nations have no interest in developing their global counterparts? This applies

mostly to trade where the efficiency and quality of goods and services is better off when

partners are on equal footing. Will industrialized nations leave developing countries to

repeat mistakes as they fall into the same struggle towards sustainability? It comes down

to whether or not we consider slave-like labor a historical mistake or as part of a

progression in developing economies. This holds grim prospects for the worlds' future

generations as it implies civilization cannot learn from past mistakes, nor escape systemic

hardship.

In an essay defending the occurrence of sweatshops, economist Murray Weidenbaum

argues that industrialization in Southeast Asia is a "transformation…not as the result of

idealism, but from changing economic circumstances."[xvi] Such beliefs rely on economic

theory and idealism that in many instances has more presence on paper than on factory

floors. Some economists argue that interfering with this "transformation" is itself

idealistic protectionism that disrupts a natural process. It's viewed as something of an

overbearing parent trying to advance a child's development by setting strict guidelines,

which actually hinder growth. These fears, mostly in libertarian economists, are of state

paternalism impeding the development of market forces. This puts their complete faith in

markets alone to get the job done right. It is a matter of faith mostly where formulas are

not mathematical certainties, as the socio-political factors of a nation affect its people

foremost.

Weidenbaum also points out the historical view found in economist Milton

Friedman's argument that his parents worked in sweatshop conditions to earn enough for

his education. This was at the last throws of America's industrial revolution and

Friedman's view holds that parents, the previous generation, endure hardship so as to bring a better life to their children and their future. If markets are nourished like a parent raises a child, essentially every parent hopes that a child will not have to make the same mistakes. Weidenbaum and Friedman are in essence defending sweatshop conditions as hardships that are a precursor to stronger development. Except that in the globalized economy there is little guidance given that goes beyond what is immediately profitable, leaving developing markets undernourished. Another libertarian voice, Johan Norberg, suggests that, "the economies of industrialized nations have passed through the transformations that the developing countries have ahead of them…they can take shortcuts and learn from our mistakes."[xvii] Here again, defenders of market forces claim a beneficial parental relationship to developing economies, but one where the parental corporate forces behind nations continue to receive abusive reports.

Historically, while development has flourished, it has in no way been a shortcut that is formative in progressing human rights or fairer trade. In the free markets these thinkers advocate for, what role have multinational corporations played in developing undernourished economies? Sure they provide the basics of industrial survival, but workers who make up the health of development are used up as means to a distant corporation's ends. Considering the running record in both labor and environmental atrocities of many multinationals like Nike, Coca-Cola, Bechtel Group, Royal Dutch Shell, Monsanto, etc., have with trade expansion after World War II, they've hardly been responsible global neighbors. The same rules and laws that ensure free markets for international trade, also protect these corporations from accountability.

Many suppliers in South Asia, the global factory of consumer growth, are not just

operating with short-term inefficiencies or temporarily bad labor conditions.

Manufacturing conditions have not gotten better so much as new factories of the same

poor standards have started up. Now with the growth of the electronics industry,

manufacturing plants that produce for tech companies like IBM, Hewlett-Packard and

Dell have been found to operate like sweatshops.[xviii] There are abuses and conditions that

any business manager could see as hindering productivity by endangering worker's

health, unless you're a slave driver. For a multinational corporation to support a factory

that essentially abuses workers is not, as they claim, bringing economic development to a

region. It is a way to take advantage of undervalued labor and cheap prices; it is a

cutthroat mentality that is detrimental to growth in the long haul.

On the surface of opening free markets it may seem that everyone is happy, with

corporations getting competitive prices while a region gets needed jobs. Though the areas

are rarely developed as long-term investments and in many cases cheap labor is used up

and left behind for worse. Once conditions at a factory improve to the point of higher

wages or workers rise up to demand them, the business they depend on from

multinationals vanish in search of the cheaper prices produced from lower wages.

Theory vs. Reality

It is understood that business must seek to minimize costs and increase profits; this

keeps shareholders invested and workers, for the most part, paid. One hopes that the end-

users of products are happy too, either with lower prices or reasonable quality. The past

mistake of sweatshop operators in the industrial revolution was the failure to see that productivity is lost in unhealthy workers. Even if not life-threatening, workers can be

demoralized by nasty conditions causing low productivity. This management ethic may

not have been fully realized until the 1950s when thinkers such as Abraham Maslow

began to introduce basic psychological needs into business practices. This view had been

developing for many years, practically by entrepreneurs like Henry Ford, but also in the

more extreme sociological theories of Max Weber or Karl Marx. Yet as with the

abolishment of Slavery, the wrongdoings persist and sweatshops continue to this day as a

prevalent business practice. This is what one scholar noted as "obstacles to self-

realization...so deeply rooted in the institutions of our economy that only profound

changes in the ownership and control of industry can create the opportunity for greater

access to meaningful work."[xix] While Friedman may be right in that forebears of an

economy endure hardship so future generations enjoy higher standards, the theory has not

followed through for many economies.

The boom of trade agreements and global industrialization that came in the years after

World War II excelled beyond what any sociological theory could keep up with. Other

more devious theories were being applied to the business world after WWII. As an

accessory to the baby boom in the U.S., huge markets opened up as the troops returned

home renewed with a thirst for the good life. For most people this meant consumer

goods, like houses, cars and all the appliances or low priced apparel one could desire.

While the influential theories of psychologist Sigmund Freud were used in marketing and

advertising since the early 20th century, the consumer dreams of the post war years were

thoroughly exploited.[xx] Another breakthrough was also happening during the 1950s as the

U.S. invested its newfound position in Asia and restructured Japan as a major trade partner. One of the largest groups of U.S. exporters, Cotton producers, found substantial

outlet in the growing Japanese apparel and textile industry.[xxi] In turn Japan's inexpensive

fabric exports found somewhat enthusiastic buyers in the West, all due mostly to U.S.

military oversight of South Asia's industrialization. Trade unionists back in the U.S.

expressed the first fears of outsourcing then, entering a stage of economic protectionism.

Meanwhile, wars waged by western nations in South Asia, the Korean War and the

Vietnam War, were driven both ideologically against communism and industrially by

capitalism. Whether the reality of the promise of industry coded as democracy realized

fruition in Southeast Asia is a matter of perspective to be debated. Tokyo, Hong Kong,

Saigon, Malaysia and others may have all undergone massive economic progress, but the

accumulated incidents of slave-like labor show market success for foreign investment

rather than a region's people.

Freeing Trade

   In the post World War II years and throughout the conflicts that followed, there was

an influx of economic investment from America into Southeast Asia. Industrial

revitalization was subsidized by the building of apparel and textile manufacturing in

formerly colonized countries like Hong Kong, Malaysia, Thailand, the Philippines,

Indonesia, Singapore, as well as Pakistan and India. Most of this economic support was

intended to feed Japanese growth and establish the primacy of "Western-Dominated,

free-world networks of trade and investment."[xxii] These were largely government-backed

initiatives, but the private sector was a powerful silent partner. Manufacturing in almost

all industries can trace some export activity to Japan's postwar boom.

The seeds of attempting a Global trade harmony go back to the failed effort of the

League of Nations, but found firm roots in the United Nations after WWII. It's worth

mentioning that the International Labour Organization (ILO) survived this transition of

world bodies, but didn't have a say in resulting trade agreements. When industrialized

nations implemented The General Agreement on Tariffs and Trade (GATT), which

formed the basis of the WTO in 1994, the initial purpose was reconstruction after the

war. In time a great deal of the trade agreements, which aimed to bring capital from

powerful nations to developing nations, created Export Processing Zones (EPZs). Known

also as Free Trade zones, the EPZs relaxed tariffs, quotas, taxation and labor laws

through the agreements, allowing multinational corporations access to cheap

manufacturing outside the realm of state control. These zones used sub-par labor wages

to attract foreign direct investment, while manufacturing could be divided up so that any

domestic production taxed only what was done within borders. For instance the Nike

Corporation could outsource part of their production in Asian EPZs and complete the

process in the U.S., paying taxes only on parts produced stateside. The Nike corporation

got it's start exporting shoe manufacturing to Japan in the 1960s, and slid down the wage

scale through Southeast Asia as Japan's pricing (wages) increased. Professor Andrew

Ross calls Nike's history a "textbook illustration of the logic of the multinational free-

trade corporation—...its enthusiastic participation in the 'race to the bottom' in countries

under authoritarian rule like Indonesia, Vietnam and China…"[xxiii] Nike and other

multinationals can argue of their impact on the increase of minimum wages in countries

like Indonesia, but this did not come until the early 1990s when it rose more than 300%

from under $1 a day to over $2 day.[xxiv] This came around the same time Nike implemented its manufacturing code of conduct in 1992 due to activism pressure, serving

another example in the late coming of many corporations to own up to an inkling of

accountability.

Early international trade agreements had a promising ring for the Asian countries that

sought a place at the table of global trade. This continues today where, as Doug Kellner

explains, "NAFTA and GATT trading standards treaties have made it even easier for

Nike and other global corporations to move production across the US border—

Consequently, Nike is able to shift around its manufacturing at will, searching for the

lowest labor costs and most easily exploitable working conditions."[xxv] To stay

competitive, corporations island-hopped around the world creating EPZs in the Pacific

Ring, stretching from Southeast Asia to South America. While not in direct opposition

with arrangements that created EPZs, a number of ILO reports reveal that the promise of

trade has deceived some regions, only to result in capital flight. This would pull the

magic carpet ride of free trade right out from under countries who had no ability to

recover. Workers were lured into EPZs, sometimes involuntarily, as factories operated

under Western investment and product orders, but after many years conditions have seen

little improvement. This game is still played out today and we are again faced with the

problem of letting economic change regress into historic mistakes. In her book on

globalization and sweatshops, Ellen Rosen draws on documented reports and historic

figures confirming, "ILO findings do not offer hope that export processing of apparel, as

it is now structured, is likely to generate stable forms of economic growth."[xxvi] Not

surprisingly it is in these zones where some of the world's worst working conditions exist, as an indifferent class of people get rich by staying adrift from the humanity of fair labor.

This is not entirely to claim the age old discontent that the rich are getting richer and the poor getting poorer. While statistics show an increasing income gap in the fastest growing economies of China, India and Brazil, free-market liberals can stack solid numbers as proof otherwise. In Johan Norberg's book *In Defense of Global Capitalism*, he offers a heavily evidenced position that "Free Trade" defies this notion of expanding income gaps. The rich may be getting richer thanks to free trade, but according to Norberg and several economists, not as fast as the poor are prospering. This is cited from studies focusing on regions that have opened up to free trade, showing that newly liberalized markets give the poor accelerated growth that actually slows gains for the already wealthy.[xxvii] This completely flips on its head the claim of anti-corporate globalization thinkers that free trade agreements create wider income gaps. Yet either side of this debate can readily admit that economic studies are massive undertakings with numerous factors that may hide realities to produce results. With such debates that are reliant on cherry-picked data the path to solutions is never clear.

This complexity is also true of understanding how Free Trade has become such a powerful motor in corporate Globalization. President Eisenhower's "Military-Industrial Complex", extends an intimate agenda of powerful nations, with trade agreements such as the GATT, and today of the WTO, NAFTA and CAFTA, serving a multitude of arrangements to meet one goal; trade liberalization. This is not a hidden agenda, but the looming question of whether free trade policies benefit all nations is what meets suspicion from critics. Since the time Merchants set sail to the open seas, trade has always sought to be freer amongst nations, especially in the building of empires. Of course the imperialist power always sought the better end of the deal and today's powers strive for no less. The philosophy of Free Trade extends back to even the 13th century where Spanish Monarchy established "free zones", an early precursor to export processing zones.[xxviii] It is a rational mode of trade to be open and free to exchange with others so as to get the best value of goods that may not have been previously accessible. Though as economies grew, and nations began to trade for what they already produced, merchants often took on a more protectionist tone. As with the 18th century British mercantilists, the philosophies of economists like Adam Smith and David Ricardo arose in opposition to mercantilism and trade protection. The extreme positions of libertarians who wanted complete autonomy from any state control grew with this Free Trade ethic. The vehicle through which this economic philosophy would excel would be capitalism.

The realities sidestepped in these grandiose theories have always been the labor that builds the towers for the idealists. To weigh the larger historical impact of trade and globalization is to question whether trade arose of the need to manage growing populations, or if trade itself spreads the ability for populating? Leaving this question to the quest of other books, the aspect that remains is that the majority of these populations have been laborers. When Karl Marx set out to expose the neglected humanity of labor in the capitalist theories rooted by Adam Smith, he was stuck in inevitable agreement with him. Future economies were both propelled and overshadowed by the understanding that "capitalism, in their view, is an economic and technological juggernaut beyond anyone's control."[xxix] The international division of labour in the capitalist system for both Marx and Smith doomed workers and would only find hope in prudent oversight of the economy. Just how Marx and Smith looked to achieve this oversight is lost in the over-analysis and continual misconception of their theories.

What remains important is that we prevent market fundamentalism from crushing the workers who enable larger philosophies of free trade to exist. Smith undoubtedly would have been opposed to state involvement in the prudent oversight or foresight he and Marx turned to for worker protection. Though a fight can only be regained on the battleground it was lost and it was governments that enacted or allowed early trade arrangements detrimental to laborers.

Choking the Tiger's Roar

The position of China today poses a dilemma for libertarians in seeing state protection as interference with functional free trade. Entire regions of China operate in Special Economic Zones[xxx] that provide export-processing arrangements within the controlled involvement of the Chinese government. At the same time, laws have been implemented in China, protecting companies from labor rights. Though their state sanctioned Free Trade zones may be better off comparatively, China still has one of the worst human rights records in the treatment of workers. In 2002 another authoritarian country North Korea, formed an isolated pocket of Free Trade zones near the Chinese border, to experiment as China is with capitalism. Just where this experiment will take China and North Korea is something yet to be seen, but initiatives lacking a basis in human rights further taint the reputation of participating multinationals.[xxxi] Economists are quick to point out the boom of China's miraculous growth in the 1990s after the government's liberalizing experiment while maintaining authoritarian rule. With the country's markets opened wide, a tsunami of foreign investment poured in and there was an exodus of youth

from the countryside into new urban factory towns. Metropolis skylines popped up and

unforeseen futures for China's next generation led them away from their family farming

roots.

As is common throughout China and all of South Asia in Bangladesh, Vietnam,

Indonesia, Thailand and others, new factory jobs were made particularly appealing to

young women and girls. While child labor problems at the focus of human rights activists

and international pressure have largely dwindled, teens and women in their early twenties

succumb to the promise of opportunity. Sometimes rural families pay fees in order to

have their children go off and work and it is true that there are often lines of

impoverished applicants waiting for a spot on the factory floor. It would seem that these

workers are taking the only opportunities that exist in that Friedman sacrifice of hardship

to pave an economy's future. There is something amiss about China's prosperity though

and as fast as foreign investment dollars grow; labor conditions seem to get worse. As the

largest producer of the world's goods, China's financial freedoms are a distraction from a

political stranglehold. Even market advocate Johan Norberg admits in his defense of Free

Trade that inequality actually increased in China with this growth.[xxxii] Developing

countries around Southeast Asia and South America can be cherry picked for data with

connections between free trade and poverty reduction. Yet China's towering example of

market liberalization is a whole new monster that overshadows harsh realities.

Perhaps in a country of such populous numbers it is inevitable that such disparities

will exist. Some may blame China's authoritarian rule for hindering the true potential of

market liberalization, but what of India? Another country whose economic prosperity is dampened by growing income gaps, but India is one of the world's oldest democracies. Under careful analysis government cannot be entirely to blame for market inefficiencies, nor is free trade the cure-all to poverty. While opening Special Economic Zones has generated jobs, income and the prospects of modern amenities for China's rural landscape, the same Global tensions exist. Workers in America and Europe have found the epitome of blame in job outsourcing to Asia, but also China's rapid climb has dumped jobs as fast as they are created. Large scale construction and manufacturing in supposedly protected zones in China face the same dilemma as countries victimized by corporate island hopping in search of lower wages. Hundreds of thousands of workers form protests in China increasingly more every year, many which are, as Andrew Ross mentions, "workers in export zones, most of them migrants in some form of indenture…"[xxxiii] Ross also points out that the political remnants of China's socialist roots with a responsibility to populism and the workers, has stalled some job loss. Ultimately, the juggernaut capitalism prophesized by Smith and Marx will not appease China's populist nostalgia.

Theory Therapy

As industrialization has shown the world over, worker satisfaction is not part of the capitalist dialogue, regardless of its importance to productivity. What is most promising though is that critics of free trade and corporate globalization are not demanding an end to capitalism. The emphasis is on Fair Trade movements and political diplomacy that can create cooperative nations so that shady export zones are not the only option. Trade most likely is the answer to peaceful coexistence on the planet and that aspect of GATT, the WTO and IMF shine through from WW II reconstruction. As historic tensions and grudges die down, the task of this generation is to take advantage of a global perspective with a long hard look around before new tensions arise.

Sociological theories unnerve the central agendas of economic theories and as people regain lives outside Freudian consumer branding, alternatives become possible. Healthy economies can flourish in free-markets, developing regions so as to best utilize their resources and benefit directly through trade. Globalization is not dependent on cheap manufacturing that feeds a frivolous consumer culture. In the untainted beliefs of market advocates, like Norberg, the true essence of free markets is people taking control over their lives and choices. As old as free market ideas may be, we have only begun to extricate valid questions from globalization's web of theory. Fair labor presents the most visible urgency and as Ellen Rosen concludes, "raises serious questions about the potential of free trade and free markets to bring economic welfare and individual freedom to workers...throughout the world."[xxxiv]

Similarly, this chapter has just dipped the surface of deeply rooted causes that have led to the new labour movements. It will hopefully show that one must reevaluate bite-sized portions of theories stretched far off their origins, such as with Smith, Marx, Weber or Friedman. History's lessons are often found beyond the textbooks and realizing that every nation has its own version to tell can enhance understanding. What lies ahead is to transcend the minute-to-minute media sphere that informs our world. This media has largely shaped the views held of sweatshops, but much slips through the cracks in that 30-second sound bite. This is mostly because history and economic theories can't be crammed into sensational stories that sell. Years of ideological imperialism dictating the market forces that shape the economies we live by just don't make it on primetime.

What's left for those who seek answers is a surprising unraveling of just how much

sweatshops play a part in the human drama of supply and demand. Once people grow

above their role as just "consumers", we come to realize that supply and demand is made

up of our own choices.

◯◯◯◯◯

[xiv] International Labor Organization Statistics Bureau, http://laborsta.ilo.org/ (Accessed 4/06) Statistics are most recent data since 2000 for the countries: Bangladesh, Brunei, Cambodia, China (Hong Kong and Macau), Japan, Korea, Malaysia, Maldives, Pakistan, Philippines, Singapore, Sri Lanka, Taiwan, Thailand and Vietnam. The included jobs are under ILO International Standard Classification Occupations (ISCO-88), Groups 6, 7,8 and 9.

[xv] "EPZ Employment Statistics" from the International Labour Organization, http://www.ilo.org/public/english/dialogue/sector/themes/epz/stats.htm (Accessed, 5/06)

[xvi] Murray Weidenbaum, "Enlightened Standards" in *Child Labor and Sweatshops*, (Greenhaven Press, 1998), pg. 27

[xvii] Johan Norberg, *In Defense of Global Capitalism*, (Cato Institute, 2003), pg. 278

[xviii] As reported by Graeme Wearden for CNET news in 2004 about a report released by the Catholic Agency for Overseas Development (www.cafod.org), article link: http://news.com.com/2100-1022_3-5148328.html (accessed 5/06). Many of the companies singled out in the report have since addressed the issue.

[xix] James Bernard Murphy, *The Moral Economy of Labor*, (Yale University Press, 1993), pg. 228

[xx] "Consumerism" An in-depth report from the BBC reveals this history of consumers and exploitative marketing tactics. The segment about Freud's Theories is included in the "Sell, Sell, Sell" discussion program, in "The Product" section of the program. http://www.bbc.co.uk/worldservice/specials/145_consumerism/page8.shtml, (Accessed, 3/06)

[xxi] Ellen Israel Rosen, *Making Sweatshops: The Globalization of the U.S. Apparel Industry*, (University of California Press, 2002), pg. 39

[xxii] Ellen Israel Rosen, *Making Sweatshops: The Globalization of the U.S. Apparel Industry*, (University of California Press, 2002), pgs 37 - 50

[xxiii] Andrew Ross, *High Profile, Low Pay: The Global Push For Fair Labor*, (W.W. Norton & Company, 2004), pg. 99.

[xxiv] Hartman & Wokutch, "Nike, Inc.: Corporate Social Responsibility and Workplace Standard Initiatives in Vietnam" in *Rising Above Sweatshops* (Praeger, 2003), pg. 147

[xxv] Douglas Kellner, *Media Spectacle*, (Routledge, 2003) pg. 81

[xxvi] Ellen Israel Rosen, *Making Sweatshops: The Globalization of The U.S. Apparel Industry*, (University of California Press, 2002), pg. 26

[xxvii] Johan Norberg, *In Defense of Global Capitalism*, (Cato Institute, 2003) pg. 135

[xxviii] "What Are EPZs?" in "Labour and Social issues relating to export processing zones" from the ILO Labour Law and Labour Relations Branch, http://www.ilo.org/public/english/dialogue/govlab/legrel/tc/epz/reports/epzrepor_w61/1_1.htm (Accessed, 5/06)

[xxix] James Bernard Murphy, *The Moral Economy of Labor*, (Yale University Press, 1993), pg. 182

[xxx] "Special Economic Zones" are exclusive Chinese Export Processing Zones that operate as

[xxxi] One example is that of the corporations Google, Yahoo and Cisco in China who were brought under U.S. congressional hearings for selling technologies to the Chinese government that allowed state surveillance, which conflicts with Western values of basic civil rights.

[xxxii] Johan Norberg, *In Defense Of Global Capitalism*, (Cato Institute, 2003), pg. 133

[xxxiii] Andrew Ross, *High Profile, Low Pay: The Global Push For Fair Labor*, (W.W. Norton & Co., 2004), pg. 142

xxxiv Ellen Israel Rosen, *Making Sweatshops: The Globalization Of the U.S. Apparel Industry,* (University of California Press, 2002), pg. 244

Chapter 3:

The Distancing Effect

"We make a terrible mistake if we think that the sweatshop workers desperate to hold on to their job are grateful to us."[xxxv] This bold statement from author Charles Derber is something many in the corporate world still have yet to embrace. These desperate workers, such as Derber describes, are sweated to the bone only to return with bare-subsistence wages in their often one-room shack homes. In China's special economic zones men and women workers share tiny dormitory rooms with up to eight people of which they must pair up for beds. Bathroom breaks are unheard of in sweatshops and workers face fines or firing in any action other than quick, silent hard labor where un-reached production quotas are met with punishment. In Bangladesh, despite existent labor laws, women workers still struggle to make up to even just 34 cents an hour.[xxxvi] In some factories, workers risk their lives in trying to create collective efforts in just asking for a few cents more.

Similarities exist in the conditions for workers in South America, South Asia and increasingly in the Middle East and Africa. Highly reminiscent of immigrant communities in the U.S., even today Sweatshops are still a reality in industrial nations. These people, our fellow human beings, are for the most part already poverty stricken and outsourced manufacturing jobs are conveniently the only option. While there may be resources in their lands valuable to agriculture and industry, international trade policies have often positioned indigenous populations out of the deal.

The workers in international manufacturing are more often women who support a family on wages barely enough for one. Sweatshop operators have always preferred women workers, not because of their dexterity or intelligence, but because they are perceived as the lesser sex. Like child laborers, women are less likely to make demands for basic rights and are easily manipulated. In a state of extreme poverty, women and children are not the only vulnerable workers and men are just as likely to be abused, but all see any job as a blessing. Organizations like United Students Against Sweatshops have focused on building a system of solidarity with workers, so not to victimize their plight. This is essential on the end of organizing workers, but there is often a first step in reaching out to workers requiring a realization that they are victims. From there, solidarity can be formed in union organizing or in building pressure on governments and corporations to reform conditions.

With a relative impact in the past twenty years, gaining much momentum in the past 10, its time for the activism to evolve. It is often said that the only job of the activist is to raise awareness of the issue, not provide solutions. That is the position those in question must take or for those with the power to create change. As much as we are all contributors to the global economy, our roles in creating a sustainable and equal one are different. The time is coming for another stage of solidarity where real solutions must first pass several hurdles. There are many who need to jump those hurdles, but consumers have the advantage to take the lead. Solutions are seeded in their ability to communicate with their dollars to corporations and governments. The ball is now in our court as consumers and we must demand a position as team players with corporations. Before this

can happen consumers must stand on common ground about how globalization will

function to provide the goods we need.

Distant Forces

As buyers of goods in the western world, consumers are displaced from the realities of

manufacturing by the marketplace. To think of the plush retail fronts or massive

selections lining the shelves would amaze the production workers, especially when they

pick up the price tags. There may me an expanding playing field with globalization, but

this may initially cause harm as well as uncertain benefits. As Charles Derber explains,

"…a key flaw of the current dominant system is that markets respond to the wants of

those with money and disregard even the most basic needs of those who do not have the

means to pay."[xxxvii] The examples of this are so wide spread that it forms a quilt of

indifference blanketing harsh realities just below the surface of industry. There are far

more dire situations, such as the AIDS epidemic in South Africa where treatment cannot

be administered due to a lack of infrastructure on the continent.[xxxviii] Not only in Africa,

but everywhere that poverty permeates human life, survival depends on being able to

afford it. Millions of children are born as victims of poverty and the system that traps

them is one of our own making. One cannot get stuck rationalizing global inequality by

saying that without industrial progress solutions for disease and poverty would not exist.

As much as that can be argued, there is no excuse in denying fellow humans a helping

hand in time of need. It's nearly ludicrous to rationalize this when the very people in need

are slaving away to make distant investors richer and products cheaper.

Though the one fault some activists may have no matter how well intentioned, is attacking executives from the on set and not engaging them in solutions. It has been mentioned that it is not the job of activist to implement solutions, but much of the anti-sweatshop movement has taken this extra step. Its' easy to feel that chance and circumstance position rich executives as a greedy nemesis to those who struggle with poverty. Still, imagine yourself a successful middle-aged entrepreneur who has worked day and night from small beginnings to build a company that employs thousands. Every time you see someone wearing or using your product, a feeling of pride and satisfaction swells inside you. Then the bloodthirsty media picks up reports from activists that you are not only responsible for, but also profiting from horrid working conditions. Suddenly you are a sweatshop perpetrator and you must question every moral fiber within you to accept or deny blame.

This is the dilemma that confronted the likes of Phil Knight (Nike), Michael Eisner (Disney) and Lee Scott (Wal-Mart) as leaders in today's economy. While the hand that strikes the abuse of impoverished workers may not come from the boardroom, it is their decisions that allow abuse to continue in factories. For years, and to this day, several corporations have ignored the need for oversight of basic worker's rights because the price is right. Denial can be a dangerous trap and if the evidence is overwhelming, which in the case of Nike, Disney and Wal-Mart it is, it is time to own up. In fortunes fueled by a form of human slavery, no amount of philanthropy can disguise the underlying greed. Andrew Ross insists that major multinationals "make donations to nonprofit causes to launder their public image, while employing workers who often toil for fifty straight hours and can still barely feed their families at the end of the week."[xxxix] Furthermore, their operations risk collapse under an economic system increasingly exposed negatively to the consumers who support it.

This system is the sea upon which captains of industry sail forging ever-greater wealth and prosperity into endless horizons of free trade. Ross also reveals a telling philosophy underlying corporate decision-making in Jack Welch, who was CEO of General Electric for 20 years. Welch saw the ideal manufacturing system as a fleet of barges that could set sail in search of the best labor markets, which of course meant the cheapest labor to exploit. According to Ross this is "an investor's fantasy and a union organizers nightmare."[xl] The problem with this fantasy is also that the world is round and companies will run the course of exploited labor, eventually crashing full circle into their own harbor. Where the world's ruthless entrepreneurs once carried slaves across oceans to maximize profit, today's ruthless entrepreneurs bring the factory to the slaves. Within the past half-century they need not fear a slave-revolt, for when a regions' labor force made demands, they just packed up and left, or replaced them. In the full circle of globalization, how long can this last if corporations are, as they often claim, building developing economies?

This is seen in Stephanie Black's documentary *Life And Debt*, exposing Jamaica's damaged relationship with the International Monetary Fund (IMF) and World Bank. Operating in one of the country's Free Trade Zones, a factory making American Brands like Tommy Hilfiger fired the Jamaican workforce when they demanded better wages and conditions. The factory owners shut the Jamaican workers out and brought in Korean workers at much lower wages. U.S. companies that contract foreign labor get the government to use taxpayer money to lobby developing countries into producing cheaper exports. In return a country like Jamaica gets needed jobs, but under conditions that allow multinationals to pressure factory operators into the lowest possible wages. This means the lowest possible operating costs, which in turn makes labor expendable and as in this case, Jamaica lost the very aspect it signed up for in a Free Trade Zone. The country may have initially gotten foreign direct investment, but once their domestic labor force is disposed, they return to draining a country's already weak social system. This is evident in a country like Haiti, where a tumultuous political culture doesn't mix well with the ruthless trade policies that companies like Disney take advantage of.

When activists turned up consumer heat under Disney's production foot in Haiti, a company spokesperson had only this to say, "We're a very small part of what is transpiring in Haiti…We have nothing to do with the operation of the government, nothing to do with the standards of living of these people."[xli] This came at a time when multinationals hadn't begun to accept responsibility for sweatshop abuses and largely still haven't. Disney, among other multinationals failed to see their impact economically on a country like Haiti, who must submit to trade policies that favor wealthier nations. Defensive companies will readily explain this as trickledown economics where if the rich continue to profit, supposedly so will the workers at the bottom of the corporate pyramid. Though this has yet to manifest at many levels, especially with the constant labor struggles seen internationally in many industries.

Unfortunately the attitude of getting the cheapest cost to make the greatest profit is part of another trickledown to consumers as well. It's one big selfish race to maximize individual self-worth with no regard for how it happens. It is a fundamental attribute of human nature to reap the most reward using the least work, but one that needs reexamining. There is a great deal of sociological studies in business that show how marketing and advertising of a company reflect what consumers wish to buy into. Taking a step back when examining how global manufacturing works can reveal a disturbing extension of this. As corporations work harder to compete for loyal consumers, they must bear a transparent, clean, operation. No one wants to buy a product that says they are contributing to a company that cares next to nothing for the people who build their products.

Consumer demand drives a need for cheaper goods on several levels including, affordable living, growing populations and technological advancement. While all these factors create the need and availability of cheap products, the question of sustainability arises. Not just if the supply chain can sustain a weak workforce, but also if industrious propagation is out pacing environmental sustainability. The story of Wal-Mart's rise to the largest corporation in the world begins with the search for the lowest possible prices. From Sam Walton driving across town in Arkansas to buy cheaper ladies underwear to their power wielding influence in outsourcing manufacturing to China. As the biggest player in the market, they set the standards of pricing that force wages down and competition must sink to their level. For activists fighting against this corporate hegemony, its' not just that Wal-Mart is the target with the biggest bull's-eye, but also that they tout a consumer loyalty to the lowest prices. Breaking this loyalty, or what some might call an addiction, is like trying to pull the suckling young off the mother-pig feeding them. Though it cannot be overlooked that consumers may initially benefit greatly from large discount chain operations, how much of it is just a false, manufactured sense of dependence.

Wal-Mart is a large piece of the puzzle in globalization, where hegemony trumps efforts to build a harmonious economy. Author Doug Rushkoff recognizes that, "the company's stiff set of competitive business practices has led it into a self-defeating cycle of corporate insensitivity and shortsightedness."[xlii] One could say that Wal-Mart's rise to the top of the supply chain forced it to reflect a common self-destructive trait in any type of imperialism. Notions that the Wal-Mart system just works and you can't beat the prices drown out the cries of activists. A company has every right to seek the lowest prices and consumers every right to demand them, but as Robert Greenwald's scathing documentary about the retail giant claims, there is a *high cost to low Price*.[xliii] The workers, both outsourced and domestic, pay the highest price as revealed by several exposés with former executives. One of which who traveled to Honduras for the company and was denied in his efforts to improve the lives of sweatshop workers there. This is a sad, but common truth and one that forces some activists to dedicate a lifetime of struggle bringing the reins of empowerment to laborers.

There are few opportunities to progress worker's rights like the one Charles Kernaghan and the National Labor Committee had with the brave Wendy Diaz from Honduras. Standing before a congressional hearing as camera's broadcast her testimony to the nation, Diaz turned the apparel supply chain inside out. Kathy Lee Gifford shed as Kernaghan said, "crocodile tears" for the sweat of children toiling for her Wal-Mart clothing line, but it didn't t take long for a mop up on that aisle of the store's image. With Nike under attack by the student movement, the face of sweatshops glaring from Diaz's determination into Gifford's smeared make-up, it looked as if things would change. Yet as economists had known for some time, the problem is so much larger than just clothing and takes more than a media circus to get the show on the road. It needs more than a

catchy slogan with Nike swooshes as frowns on happy faces or video's where Mickey

Mouse takes you on a Haitian sweatshop tour. These efforts turn the tables with creative

brilliance to raise awareness, but in the end it is consumer choice that turns the tides.

Means and Ends

The first hurdle awaits consumers is the ideology that has guided the latter half of the

20th century. It realized its fruition with tremendous foresight, for better or worse, from

classical economists. So goes the totem that as companies and individuals maximize

profit the "invisible hand" of the market benefits all. This in theory is morally sustainable

when all those who act within the market have the best intentions in all they do, a

utilitarian mentality of greatest good for the greatest number. Naomi Klein reveals a

current strain of thought from today's economists, "…who spin the mounting revelations

of corporate abuse, claiming that sweatshops are not a sign of eroded rights but a signal

that prosperity is just around the corner."[xliv] Outside the spin, what happens when greed

and ruthless competition possess those who shape the markets? It encourages the same

from, employees, consumers and all who are seemingly cooperating, creating a chasm

between the ends and means of lives inevitably linked.

Corporations today rarely operate within means that are consistent with their

advertised ends. Of course, operations like Wal-Mart may be that exception as their end

is to supply the lowest prices and they use whatever means necessary, including

providing the lowest wages. J.B. Murphy analyzes this further within Adam Smith's

Invisible Hand and the "displacement of goals." From the theory's source, Smith himself understood that the Invisible Hand does not work if the end goal is displaced from the means of getting there.[xlv] A corporation functions as an entity that can provide profit for all it's associates and partners. This means executives, stockholders, work force, contractors, and eventually its customers. The desire of more market share and greater profit is a powerful force in successful business operations, as rational people maximize their own worth. This extends to our role as consumers when we seek the best bargain or invest in purchases that increase personal wealth. Though, when individuals inside a corporation and the customers that support them engage in short-term advantages, those means can sway from the original end.

All of this stems from the encompassing influence of Aristotle who taught that our actions should be done as ends in the means itself. This profoundly simple ethic is completely detached from business today as almost all of the capitalist enterprises produce goods and services that are a means to another end; profit. As we will see this is something pervasive through our entire economic system, which implies a disturbing lack of moral foundation in everyday life for many. Though first, its necessary to realize something the philosopher Immanuel Kant expanded on in means and ends. That being his *categorical imperative*, which essentially instructs to not treat others as a means to another end, or simply don't use people for your own gain. Again, this moral guidance or golden rule is something entirely vacant in the application of the most influential economic theory. It has maintained more of what is called an economic golden rule, where the ones with the gold are the ones who rule. As has been argued, Smith's Invisible Hand is meant so that we can all use each other as means to other ends, but if the ultimate end is not the best possible life for all, the system carries infectious greed. Though what drove Smith's work has been stretched far from his original intention. As Levitt and Dubner note in their book *Freakonomics,* "(Smith) strove to be a moralist and, in doing so, became an economist."[xlvi]

Manufactured Illusions

What is probably the most prescribed piece of advice given regarding the workplace – to love what you do, is rooted in this ethic of means and ends. While people may not always be able to make a living out of what they are passionate about, they can still take pleasure in their work. The Global economy does not permit that this is possible for all and pretending that free markets are the magic to make it happen is an illusion. Morality is not part of these economic equations, as much as they are believed to bring prosperity to poverty. A corporation may claim to have infused a region with jobs by outsourcing manufacturing, but this is not their original intent. The end goal is the obtainment of lower wages so reduced operating costs will increase profits; the means to this often come as labor exploitation. As a supply chain interplays the motives and wants of entrepreneurs, workers and consumers alike, we may unwittingly partake in means we don't agree with. This does not put customers at fault for multinationals using sweatshops, only that we as consumers are part of a complex problem.

Complex, yet quite simple in some ways, our insatiable consumer desires are due a certain level of humility. In the cartoon that spares no one for a laugh, *South Park* creators Trey Parker and Matt Stone dedicated an entire episode to their little mountain town's obsession with Wal-Mart. Absurd antics and poop jokes aside, Parker and Stone nail the crux of the matter when the heroic kids confront the evil forces of Bentonville,

Arkansas. What they find is a mirror and thus the show proclaims that Wal-Mart is you.
The driving force behind cheaper prices is that we continue to buy more stuff. It hardly
matters that you can buy 4 t-shirts for under $20 because Wal-Mart more than proves we
consume more in the illusion of it costing less.[xlvii] Consumers are finicky loose ends to
the many supply chains that try to tie down the world's markets. Some have the benefit of
not just demand but necessity like the Food Industry and Oil companies, perhaps even
textile companies. These industries can often reveal the more corrupt and deceptive
practices in maintaining that necessity.

The mass consumption of much retail goods like clothes, toys, electronics or home
accessories, is hardly necessity most the time and a manufactured demand. This is why
companies in both retail and production constantly push for the cheapest prices, because
that is the point of sale for most people. Consumers know this—You know this and some
are still led to believe their purchases are putting food on the table for distant or domestic
workers. Even if consumers want to buy "Made In the USA" products, the great Saipan
scam shows that not even that can be trusted. The reality is that ethical consumers, a
small but growing sector, do not intuitively buy goods because it is ethical or moral. It
takes a great effort to think about the impact of purchases when standing in the store and
the first drivers are usually, price, quality and style. Poking fun at how easily consumers
fall for bargains eases the passage of what is eventually hard to swallow. Essentially, that
the convenience of those prices come from, among other detrimental practices, the further
oppression of workers deserving of a living wage.

One thing that is for sure in the dynamics of an invisible hand is that we've come to
coldly expect others to serve our own ends. In a consumer culture, we have gone way

beyond purchasing things for their intrinsic value and buy based on manufactured desires to feed a status quo. In turn corporations use these desires as a means to profit, not for the sake of providing the goods themselves. What is lost is an element of trade where there is pride in the work and true enjoyment in a balanced exchange, not just rampant consumption. To make this cold exchange a continuing realty, consumers are saturated with advertising to maintain supply to a false demand. Corporations then resort to using desperate conditions of foreign workers as a means to reduced overhead. At the helm of it all from behind closed doors of corporations and trade organizations policies are shaped for profitable domination. With this, as Hartman, Arnold and Waddock add, "free market theory suggests that people work in sweatshop conditions because it is the most rational means available to them for furthering their own ends."[xlviii] This combination of needs from desperate workers and corporations seeking low labor costs would seem to find a common ground.

So in the end everyone gets used, but achieves their desired goals, right? Wrong, the system does not run on a level playing field, its designed to feed upward into corporate profits. Only if labor costs can be reduced substantially and six figure salaries maintained do consumers get a fair price-even just some of the time. One of the biggest pushes behind the expansion of Free Trade policies is more people get access to more affordable goods. While this is evident is some instances, in the case of apparel especially, Rosen's research shows "that the benefit consumers derive from this is minimal."[xlix] Where workers give all they've got, blood, sweat and tears to make anything, their ends of survival do not mesh with a system that drains their energy for profit.

The problem will persist no matter how much consumers know, as long as there is inaction. We live in a world with options so vast that it has spoiled a vast majority of people with choices they are not willing to, nor should they abandon. Collective change is slow especially when it seems to go against the grains of progress, but in fact taking a step back can actually be an act towards greater progress. When we learn that certain brands of shoes or clothes are manufactured in horrid conditions, it opens the potential to question all brands and products. Manufactured goods do not have a trusted Fair Trade label like coffee beans and chocolate, which is also something most are skeptical of. Without the chance to have workers hand you goods personally with their own health in sight as a seal of approval, it is a leap of faith. Consumers buying fair trade goods are in a sense moonwalking in the world's markets. There is no gravity to the markets, so each small step, or purchase is a leap of faith and they must go to great distances for humankind. Back on the markets with gravity, others still question the point of buying fair-trade goods. Why, they ask, go all the way to the moon just to get a pair shoes?

ooooo

[xxxv] Charles Derber, *People Before Profit: The New Globalization in an Age of Terror, Big Money, and Economic Crisis*, (St. Martin's Press, 2003), pg 47

[xxxvi] Charles Derber, *People Before Profit: The New Globalization in an Age of Terror, Big Money, and Economic Crisis*, (St. Martin's Press, 2003), pg. 251

[xxxvii] International Forum On Globalization, "Ten Principles for Sustainable Societies", by Charles Derber in *Alternatives to Economic Globalization: A Better World is Possible*, (Berrett-Koehler, 2002), pg. 74-75

[xxxviii] An in depth overview of the problem with infrastructure and the AIDS virus is seen in the PBS Frontline Documentary that is available for viewing on line. http://www.pbs.org/wgbh/pages/frontline/aids/

[xxxix] Andrew Ross, *Low Pay, High Profile: The Global Push for Fair Labor*, (The New Press, 2004), pg. 37

[xl] Andrew Ross, *Low Pay, High Profile: The Global Push for Fair Labor*, (The New Press, 2004), pg. 16

[xli] Quote from Kevin Danaher and Jason Dove Mark's book *Insurrection: The Citizen Challenge to Corporate Power* (Routledge, 2003) pg. 79

[xlii] Douglas Rushkoff, *Get Back In The Box: Innovation from the Inside Out*, (HarperCollins, 2005), pg. 207

[xliii] *Wal-Mart: the High Cost of Low Price*, Dir. Robert Greenwald, Brave New Films, 2005.

[xliv] Naomi Klein, *No Logo: No Space, No Choice, No Jobs*, (Picador, 2002), pg 228

[xlv] James Bernard Murphy, *The Moral Economy of Labor*, (Yale University Press, 1993), pgs 179-181

[xlvi] Steven D. Levitt and Stephen J. Dubner, *Freakonomics: A Rogue Economist Explores The Hidden Side Of Everything*, (William Morrow, 2005), pg. 14

[xlvii] In the PBS Frontline Documentary *Is Wal-Mart Good For America*, a former regional manager of the store, Jon Lehman, gave a revealing interview about how Wal-Mart creates the illusion of low prices, forces management to cut costs and bullies manufactures into the lowest prices. http://www.pbs.org/wgbh/pages/frontline/shows/walmart/interviews/lehman.html (Accessed 6/06)

[xlviii] Laura Pr. Hartman, Denis G. Arnold and Sandra Waddock, "Rising above Sweatshops: An Introduction to the Text and to the Issues", in *Rising Above Sweatshops: Innovative Approaches to Global Labor Challenges*, ed. Hartman, Arnold and Wokutch, (Praeger, 2003), pg. 3

[xlix] Ellen Israel Rosen, *Making Sweatshops: The Globalization of the U.S. Apparel Industry*, (University of California Press, 2002), pg. 221

Chapter 4

Economic Solutions

Activism in ending sweatshops has been primarily directed at South Asian and South American countries. Following the money trail, wherever capital lands offshore, it could lead to the chained gates of sweatshop factories. As mentioned throughout the book, the trail of capital has been leading to the Middle East and Africa, in countries like Jordan and Kenya. Sociologist Jonathan London notes this capital flight in his explanation of the race to the bottom in the contests for economic development. "U.S. provisions doubling the allowable importation of textiles from Africa have occasioned a migration of East Asian capital to African countries." In the ongoing race to the bottom, "the countries with the most exploited labor "win" investment from increasingly footloose global capital."[1] Capital moves to regions where supposed development can take place and free trade zones allow companies to operate outside of their own nation's regulation. It is easy to shrug off the term developing market or economy when it seems companies are just after labor to exploit. There is no true sense of development as far infrastructure or community happening and standards of labor often stagnate well below acceptable conditions.

Part of the problem with isolating individual companies in using sweatshops is they migrate with capital to developing markets. By developing markets, this entails countries that have lessened their labor and environmental laws along with opening tariffs and trade barriers to attract the capital. Companies cannot be pinned to a specific country, as a

factory in Bangladesh could be manufacturing for the Gap, Victoria's Secret and Target

while simultaneously those labels are found on an assembly line in Mexico. Concentrated

activism efforts, such as with Nike in Indonesia, can generate greater awareness and

encourage standards of change. Though such a process is a bit like squeezing a balloon;

once you squeeze one area the air is just going to pop up in another. Even though we

don't want to be left bursting the balloon of trade altogether, many areas of reform are in

need.

Behind tainted company brand labels is a much larger dilemma of capital flight

directed by investors who second-guess markets. Free markets open up the trade winds

for the locust of capital that seeks abundance in lower labor costs. Recently in 2006, this

started to occur in Bangladesh, where millions of garment workers depend on the apparel

industry as a main component of the country's exports. The garment districts of

Bangladesh's cities are ripe with collapsing factories and workplace hazards posing fatal

risks to a workforce already without living wages.[li] Such conditions may not be

something worth holding on to, but their developing economy is threatened by

investment flight to namely the country of Jordan. The National Labor Committee

followed the money and their findings prompted international exposure of Chinese and

Bangladeshi workers in Jordanian sweatshops.[lii] Among other things, the ordeal is being

considered under the U.S. Department of State's definition of human trafficking.

In the fall of 2006, over 1,200 Bangladeshi garment workers held peaceful protests at

government offices and in strikes at factories in Jordan.[liii] The fact that workers were

adamantly protesting their rights counters the argument that without sweatshop labor immigrants or even poor nationals in underprivileged areas would be unemployed and in worse situations. These Bangladeshi nationals obviously feel cheated and abused at the hands of the factories in Jordan, sponsored by American companies like Calvin Klein, Gap, Inc and Levi Strauss. Workers are deported, their passports confiscated, and denied payment like disposable commodities with no rights, but they have begun to demand what they are worth. This is beyond national identity and wages, as workers are demanding a stop to routine violence and the mandatory 15-hour shifts; this about their freedom from enslavement.

The women garment workers from Bangladesh have perhaps the most recognized sweatshop struggles in the global economy. This is not only due to the density of the districts with millions working in some of the worst factory conditions worldwide, but it is also their voice. Author Tricia Gates Brown relays the implications of life spent "dramatically compromised," by a downward spiral of working existence. With reports of up to 20-hour shifts, only to sleep on the factory floors, sickness is rampant in an industry that does not know the meaning of healthcare. The women miss their children's entire upbringing and can leave little opportunity for them, as the next generations go on to work in the same conditions.[liv] Despite these oppressive conditions bore in the name of free trade and development, the women (and men) have managed to get their story heard. C. Derber shows that, "…they do want to keep the jobs in Bangladesh, but are passionately organizing to make the factories respect international and Bangladeshi labor law."[lv] That these garment workers are joined by NGOs, activists, student movements and labor groups from around the world, is further testament to the challenge ahead. The

Bangladeshi's voices are a loud beam of light, but yet the wheels of trade agenda push them further into the shadows of globalization.

What allows this rising exodus of labor to start is the same thing that permits what Johan Norberg refers to Bangladesh in South Asia as, "an island of free trade in an otherwise protectionist economy."[lvi] Liberal economists may also claim that aside from the risk of capital flight, freer markets are evidenced as recovering faster. It's as simple as easy come, easy go, but there is no investment in the people of these regions, it's limited to the products and that labor remains cut-rate. The same mentality existed when free trade first began and exacerbated the slave trade. By countries opening up free trade zones today the intention is a hope for foreign investment, but in reality it is clever trade manipulation that takes advantage of a duty free market.

With this a worker must leave her already hazardous, poverty wage job in one area to even more abusive conditions in another. She may travel freely as does the investment, but it is hardly a sense of freedom. Perhaps over time conditions may improve enough in such countries to support a healthier workforce, just as the claims of some Asian countries' ups and downs into stable economies. As *New York Times* writer Tina Rosenberg notes in her critiques of Free Trade, the WTO and IMF rules by which East Asia successfully entered globalization are rigged. The trade paths that countries like South Korea, Taiwan and Malaysia took more than twenty years ago, have since been manipulated in policy. [lvii] This is evidenced by several South American countries that have tried following those rules, but without seeing the promised growth rates from trade agreements. In fact, C. Derber's claim, evidenced by the World Bank and IMF, stumps the claim of growth rates championed in globalization. He notes that economic growth in

Africa, Latin American and other parts of the Third World have slowed to "a snails pace."[lviii] One result has been the uproar from populist movements in Venezuela, Brazil and Bolivia with a growing opposition to free trade policy rising from the streets into the political sphere. The North American Free Trade Agreements (NAFTA) and Central American Free Trade agreements (CAFTA) are underlined as "free" on paper, but the actual agreements are a mesh of hypocrisy in protections and limitations against developing nations without economic clout.

As J.D. London also points out, "South Korea and Taiwan have registered such improvements in living standards…that they have shed their developing country label." Furthering that argument for globalization's success, the World Bank found that the only high growth rates of the Third World were in Southeast Asia.[lix] Therein lies a larger problem though and additional to Rosenberg's claims of a free trade fix, London goes on to note disparate evidence. He adds, "…other developing countries remain deeply mired in poverty…as mere containers of surplus population, missing out on the benefits of global economic growth while incurring many of its costs."[lx] The combined distress of South Asian and South American labor struggles should be of immediate concern to economic stability. The growing income gaps and instability can be largely avoided by eliminating or redesigning polices that hinder improvement.

Lessons Learned

Within efforts such as the Organization of Economic Co-operation and Development (OECD) the potential to help developing countries transition may be better utilized. OECD's members are industrialized nations that can assist developing countries in economic growth. This can be achieved in the philosophy that the OECD promotes a free

exchange of information without the legality of rules and policy. Something that can

come from this exchange is that the extremes of free trade have been felt in globalization

in several areas. There have been extremes in both successful free trade implementation

and also in detrimental exploitation due to bad trade policies. An organization of the

OECD's capacity can even just informally dispel commonly held notions about

globalization. For instance, OECD research has shown, "the view which argues that low-

standard countries will enjoy gains in export market shares to the detriment of high-

standard countries appears to lack solid empirical support."[lxi] Since those findings in

1996, still no proof exists in the notion that factories that raise labor standards will

competitively lose out to sweatshops. There is resistance enough to see that a balance

must be struck in the thinking that shapes trade agreements, offering some form of

protection. If handled cautiously and ethically, we do not have to end up in a world full or

tariffs.

   Such a world is where classical theories arose from Adam Smith, David Ricardo and

John Stuart Mill, who reacted to mercantilists. Today's confused policies of free wielding

trade expansion are the economic rebound from that fear of protectionism. Though as this

book has argued, their liberalizing philosophies have reached extremities stretched

beyond their meaning. A more efficient system is needed where trade is indigenous in

production, fair in export and import, but flows freely. As the utopian mantra of re-

imagining globalization prophesizes, "another world is possible." This brand new hope is

a popular slogan at organized social movement events like the World Social Forum.

Sociologist Manuel Castells' points out that, "the emphasis is on "another," and the possibility can only emerge through the radicalization of the struggle."[lxii] The route of

trade we are on is headed towards conflicts over resources and people submitting to

exploitation to ensure a livelihood. Within re-imagining the bigger picture of

globalization can come the action of local solutions. Capitalism is so ubiquitous in the

modern economy that it will inevitably be a part of any form of exchange that develops.

The greater change will be, and needs to be in how trade policy deals with the world

as it is now. Agreements within world bodies need to reflect an attempt towards equality

in a disproportionate global economy. They have for too long relied upon theories that

blindly predict the advent of market forces as being an equalizer. Free Trade is a beautiful

economic theory when properly applied in historical context, but too much a convenient

code of doublespeak that hides economic inequalities.

Under the guise of free trade is a dangerous trend that John Cavanagh says is

essentially that, "corporations have increasingly high jacked the agendas of government

and of public institutions."[lxiii] He primarily addresses the power manipulation by

corporations of the WTO, World Bank and IMF. Cavanagh also sees that it is the failing

of civic attention that has partially caused this shift of power. The civic sectors of society

have let the potential of institutions created for equality, fall into the agendas of corporate

expansion and monopoly dreams. Over many years we have ignored their influence and

allowed elected politicians to ease the passing of trade agreements bent in favor of

corporate power. In the advocacy for trade reform by Lori Wallach, free trade has little

to do with trade at all. Since the World Trade Organization was started, Wallach chastises

the group's illusions where instead of promoting "free trade" they "established a 20-year

monopoly marketing right for patent holders…calculated to cost U.S. consumers $6 billion."[lxiv] As such ulterior motives come to light in the public mind, the illusion of trade doctrine is awaking to a world of harsh realities. That may be why the phrase "fair trade" is finding resonance with critical voices even from within world organizations.[lxv] These institutions are at the point where it becomes necessary in any form of social progress to sometimes break the boundaries of language and find new meaning in old sayings.

Another word in the global trade debate that won't die is sweatshop, nor has it lost much of its meaning. Some may see sweatshops as a relative matter, where they are better than existing conditions in impoverished regions. A moral imperative is at the basis of such an argument, but needs to follow through and stitch the gap between development assistance and global trade. The other side of the imperative arguing for labor rights is often criticized as lofty standards from western idealists. Such claims are brushed aside as idealists who know nothing of the conditions in which sweatshop workers live. Much of the criticism comes after factories or trade agreements have been ineffective for several years, long past the benchmark of development. How can this truly be considered economic progress when it cannot advance the people whose hands build the developing economy?

Reaction vs. Action

There is a vast void in the actions between critics of globalization and those who direct its positions, but not in what they both say. While activists protest for human rights of international workers, economic leaders disadvantage them in trade policy, as both sides claim to be advancing labor development. Part of this discrepancy comes from a great misconception that we must react to market forces and countries are powerless to change trade circumstances. What this really means is that developing nations must react, powerless to the one-side decisions of industrialized superpowers. Activists want to empower workers to standup for their rights, while market fundamentalists deny this in claims that the marketplace dictates wages. Critics of such, like Charles Kernaghan and Andrew Ross have pointed this out as absurd when trying to explain the poverty wages in Haitian workers making Disney clothes and the 325,000 times more CEO Michael Eisner makes per hour.[lxvi] The workers have fallen into the void between these two debating camps, with activists trying to pull them up on one side and misguided trade polices holding them down on the other.

This space between activism and corporately lobbied trade policy is a place made easily harmonious. Workers want their employers to profit just as much as factories want their buyers to profit. This is part of the dilemma in consumer boycotts of certain brands where factories lose orders. The larger influence of trade policy in globalization allows the belief that factories cannot improve standards and still offer the attraction of cheap labor. As much as free trade is something everyone wants in theory, it is not an excuse to take advantage of people who aren't in a position to demand rights. To put it another way, as Senator Byron Dorgan (D-ND) wrote in a letter to the editor in *The New York Times*, "Sweatshops are the problem, not the solution."[lxvii] This counters the common held belief of economists who encourage free trade because it generates jobs in depressed areas.[lxviii] The task now is to stop taking advantage of these regions and put true development in so-called developing nations.

Directing economic investment so that it establishes a working relationship with development assistance is a solution, as broad as it may be. Considering Norberg's evidence of Foreign Direct Investment (FDI), which has spiked more than 125 Billion

(U.S.) dollars since 1989, which is more than all the development assistance given

worldwide in 50 years.[lxix] Trade is the answer in this sense, but must be a responsible

liberalization where capital flows freely without leaving workers for broke. Taking the

numbers of FDI and applying it to an economically profitable program of development is

what Middle-Eastern and African countries need, not the same mistakes from South Asia

and South America. This will require idealists who want to see development happen and

have enough hope or faith that people will grow a productive economy with assistance.

As it is now, unless corporate culture makes a drastic transition to altruism, regions

may need to prove economic potential as seen in the case of India. In their economic

climb several of the nation's entrepreneurs have called for a stronger investment into the

country's infrastructure. This can ensure that a country's development goes beyond being

just another zone for cut-rate labor and offshore investment embezzlement.[lxx] While it

may not be considered embezzlement by law, there is no sense in mincing words because

what Globalization needs is words that speak truth to power.

Infrastructure investment can enable cities to grow efficient trade systems in several

industries. No matter how small the investment, it can extend to education and agriculture

to build a sustainable future for the markets. This prevents the problem of rigid

opposition to countries where sweatshop labor is often the better option for women and

children. They either may fall prey to prostitution rings or be limited to hard agricultural

labor for work. If education programs are put in place it is a start to the otherwise

imprisoned existence of many factory workers. This is how factory owners, executives

and economists rationalize sweatshops; as stepping-stones, but do so without substantial evidence of free market futures. As mentioned in chapter 2, they usually turn to China,

but that experiment is revealing some interesting, paradoxical results, with a labor

upheaval on the cusp.

Radical Changes

    With unprecedented access to a Chinese factory, producer/director David Redmon

revealed a rarely seen inner-world of factory owner Roger Wong.[lxxi] In one segment of

the film *Mardi Gras: Made In China,* their discussion explores how younger generations

have no more loyalty and lives only for the materialistic moment. Redmon takes it up a

notch to say that this is caused by the larger influence of capitalism's depersonalization of

society. A potent realization and one such younger generations are thinking about

changing, but are limited in their actions. It comes as creative expressions of realizing the

inherent contradictions of governments and corporations that talk democracy, but support

totalitarian economic control. It comes in protests that exercise immense power when

united with the workers who feel the greatest effects of globalization. Yet, these powers

are rattling the gates outside those who hold the keys to resources and a system

dependent on capital. If investment can be channeled into development assistance, these

young energies have the potential to imagine that another world is possible.

    Part of initiating real economic solutions has triggered the evolution of some from

activist to pro-activist, as they try to implement real ideas. This starts with recognizing

the problem, then pressuring the origins of the problem and empowering the people in

need of solutions. This has partly seen fruition in the talks at the G8 summit about Debt

Cancellation for third world countries. These national debts have been a cruel double-blow to countries already facing entrapped trade agreements. There are three key players

in the conflicting nature of global trade agreements that can be reevaluated cooperatively.

Organized from J.D. London's research they are:

1. "WTO Leaderships has been outwardly hostile to the idea of incorporating

   labor standards into trade agreements."

2. "The World Bank...encourages developing countries to pursue labor

   intensive manufacturing industries...in which global labor challenges are

   the most acute."

3. "The ILO... has historically seen its efforts to promote the welfare of

   workers marginalized by the economic strategies promoted by WTO and

   World Bank, not to mention state agencies themselves."[lxxii]

The discrepancies in how these institutions operate work to the advantage of certain

factions of corporate globalization. Mostly in that multinationals avoid protectionist

intervention from United Nations bodies like the ILO or governments. On the surface the

stated goals of the WTO, World Bank and ILO unilaterally advocate for increased trade

that stimulates developed and developing economies, where labor should naturally

benefit.

Something that Professor Robin Broad has suggested is not to rely upon old systems

that have failed development, such as the World Bank and IMF. She instead advocates

tipping the scales of influence towards an organization like the ILO to balance out years

of policy and trade damaging the culture of workers rights.[lxxiii] Creating a more efficient

workforce is always on the agenda for corporations and economists, but research to do so

cannot be extended without social accountability. Furthermore, research can explore the larger connections of globalization by focusing on the commodity and its value socially, economically, environmentally and in development. Political philosopher Michael Hardt explained this in the example of the film *Mardi Gras: Made In China.* By exploring a single object, like Mardi Gras beads, "reveals a world of connections and a whole world of living labor."[lxxiv] This is something Pietra Rivoli accomplishes in her book *The Travels of a T-shirt in the Global Economy,* and comes to many of the intuitive conclusions activists have been raging about.[lxxv] What both the *Mardi Gras* film and *Travels of a T-shirt...*book share is an ability to cross over lines of socio-economic research into cultural understanding. They explore, what M. Hardt calls dead objects, and give them meaning in the context of everyday life. What can amount from such studies is the many levels involved in production and trade and their impact beyond the supply chain and stock dividends.

The subversive policies that hinder growth mutually beneficial to workers, entrepreneurs and consumers, can be disposed without sacrificing profit or exponential trade. Reaching for economic solutions to end sweatshops means action in the sense of re-harmonizing the conversation and creating communication. Critics of policy within globalization need to work towards defining the radical changes needed in ways that can be transitioned realistically. By responding to this, the power elite and world bodies under criticism will in turn encourage a democratic process that encourages innovation, productivity and commitment.

☺☺☺☺☺

[i] Jonathan D. London, "The Economic Contest: Grounding Discussion of Economic Change and labor in Developing Countries" in *Rising Above Sweatshops: Innovative Approaches to Global Labor Challenges,* ed. Hartman, Arnold, Wokutch, (Praeger, 2003), pg. 59

[li] Farzin Mojtabai and Jason Cangialosi, "Bangladesh Garment District Under Fire", http://www.associatedcontent.com/article/25227/bangladesh_garment_district_under_fire.html, (Accessed 4/06)

[lii] "Jordan Accused of Harboring Sweatshop Factories", Narr. Kristen Gillespie, *Morning Edition,* NPR-National Public Radio, June 26th, 2006, http://www.npr.org/templates/story/story.php?storyId=5510902, (Accessed, 6/06)

[liii] National Labor Committee Report, "U.S. Jordan Free Trade Agreement Descends into Human Trafficking", http://www.nlcnet.org/live/article.php?id=10, (accessed 10/06)

[liv] Tricia Gates Brown, *Free People,* (Xlibris, 2004), pg. 84.

[lv] Charles Derber, *People Before Profit,* (Picador, 2003) pg. 228

[lvi] Johan Norberg, *In Defense of Global Capitalism,* (Cato Institute, 2003) pg. 223

[lvii] "Saving Globalization From Itself: Does Free Trade Need Reform?", Narr. Tom Ashbrook, *On Point,* WBUR, Boston, August 20th, 2002, http://www.onpointradio.org/shows/2002/08/20020820_a_main.asp (Accessed, 4/06)

[lviii] Charles Derber, *People Before Profit,* (Picador, 2003) pg. 86

[lix] Charles Derber, *People Before Profit,* (Picador, 2003) pg. 86

[lx] Jonathan D. London, "The Economic Contest: Grounding Discussion of Economic Change and labor in Developing Countries" in *Rising Above Sweatshops: Innovative Approaches to Global Labor Challenges,* ed. Hartman, Arnold, Wokutch, (Praeger, 2003), pg. 52

[lxi] Douglas A. Irwin, *Free Trade Under Fire,* (Princeton University Press, 2002), pg. 214

[lxii] Manuel Castells, *Power of Identity,* (Blackwell Publishing, 2004), pg. 154.

[lxiii] "Corporate Takeover", an interview with John Cavanagh on www.bigpicture.tv, http://bigpicture.tv/index.php?id=61&cat=trade&a=142

[lxiv] Lori Wallach and Michelle Sforza, *The WTO: Five Years of Reasons to Resist Corporate Globalization,* (Seven Stories Press, 1999) pg. 48

[lxv] Economist Joseph Stiglitz and Development expert Andrew Charlton released an Initiative for Policy Dialogue book in 2006 titled, *Fair Trade for All: How Trade Can Promote Development,* (Oxford University Press, 2006)

[lxvi] Andrew Ross, *Low Pay High Profile: The Global Push for Fair Labor,* (The New Press, 2002), pg. 18

[lxvii] Sen. Byron L. Dorgan, Letter, "Sweatshops in Africa? Consider the Case of Jordan", *The New York Times,* June 12th 2006, http://www.globalexchange.org/campaigns/sweatshops/3961.html (Accessed, 6/06)

[lxviii] Harvard Economist Jeffrey Sachs is often attributed a quote to this claim saying, "My concern is not that there are too many sweatshops, but that there are too few."

[lxix] Johan Norberg, *In Defense of Global Capitalism,* (Cato Institute, 2003), pgs 245-246

[lxx] William Brittain-Catlin's book *Offshore: The Dark Side of the Global Economy* explores how corporations like Wal-Mart, Enron, Citigroup and others hide investments free of tax regulation.

[lxxi] *Mardi Gras: Made In China,* Dir. David Redmon, Carnivalesque Films, 2006

[lxxii] Jonathan D. London, "The Economic Contest: Grounding Discussion of Economic Change and labor in Developing Countries" in *Rising Above Sweatshops: Innovative Approaches to Global Labor Challenges,* ed. Hartman, Arnold, Wokutch, (Praeger, 2003), pg. 67.

[lxxiii] Robin Broad, *A Better Mousetrap?* in *Can We Put An End To Sweatshops?* (Beacon Press, 2001) pg 46-47.

[lxxiv] Michael Hardt speaking in an un-released interview from the film *Mardi Gras: Made In China,* Dir. David Redmon, Carnivalesque Films, 2006. The Interview is featured on the film's extended features of the DVD release.

[lxxv] Pietra Rivoli, *The Travels of a T-shirt in the Global Economy: An Economist Examines the Markets, Power and Politics of World Trade* (John Wiley & Sons, 2005)

Chapter 5

Corporate Solutions

In all industrialized nations, corporations are as common as the clothes on our backs. The growing paranoia that corporate culture will sponsor every morsel of life is something as old as industrialization itself, but in that sense well rooted. People, on a larger social level look at each other more and more as just consumer statistics, something academically noted, but not taken seriously enough. What we wear, eat and drive is a form of our social currency. This extends to the working world, where labor becomes only the means to the goods we consume, not the backbone of the economy it is. The corporate machine is obviously at fault here, but the responsibility also rests on individuals living in a democratic society to be agents of change.

The choice in our world today is becoming not if corporations will run the world, but how they will do so. Will it be a world branded by mega-conglomerates, something like a MicroSanto-ShellMart? Or will the search for the perfect monopoly collapse the economy causing even less small business start-ups? Beneath this apparent ultimatum are the everyday operations of the worlds' largest employers. No, not Wal-Mart, but the thousands of small, independent, family run businesses that are, by choice or not, community supported.[lxxvi] Even if accumulation of profits is not in the corner of small business, the far greater force of labor's masses is there. As the largest employers in the

U.S, small businesses carry the weight of the nation's labor, but lose out to the market-share dominated by multinational operations.

The fear of corporate domination is often misdirected, as Johan Norberg reveals; the employment share of the largest corporations is dwindling and he refutes the claim that multinationals have GDPs larger than small countries.[lxxvii] Still, such figures can be constructed from sales numbers of the world's largest companies when compared to poorer countries. For instance C. Derber shows that the sales of the top corporations are greater than many countries GDP, including General Motors being larger than Denmark, Wal-Mart being bigger than Poland, and Phillip Morris' sales being greater than that of 148 countries.[lxxviii] Statistical battles aside, economies free of multinational hegemony are not only possible, but also a solution in monitoring goods free of sweatshops. One reason is it may be easier for a small, specialized company to monitor their suppliers.

This can exist on a scale of local levels where companies are in tune with community more than just consumers. They can also connect with suppliers at a domestic level and more efficiently monitor the larger global commodity chains. While it may seem that the innovative product tracking of a Wal-Mart or fashion retailer like Zara is mastering this, it also always a way for negligence to go overlooked. In respect to labor, large companies make centralized decisions and cannot be genuinely attune to the needs of its workforce. This encourages a small segment of consumers to live by the ideal of supporting small, independent business in everything from farms to filmmakers. For such a system to exist successfully requires huge shifts in economic research to study commodities, such as mentioned in chapter 4, for their impact to be fully realized. Labor also has an advantage with small business, as the larger a corporation is the less likely they are to be pressured by competition to improve standards. Overall, C. Derber mentions of small businesses

that, "economic theory suggests they have many competitive advantages over giant

firms."[lxxix]

Internal Initiatives

From a theoretical economic standpoint, it is a tremendous task explaining how

alternative globalization is possible. This is because the changes so radically redefine the

corporate culture dominating the mentality of world leaders. Protecting the rights of

workers and promoting fair trade rings alarms of paternalism in ears of the free market

choir. In considering larger multinationals that oversee their own initiatives in improving

labor conditions, while it is not the transparency demanded by critics, it encourages

progress. The mounting pressure of activism has initiated several systems of

improvement in developing nations funded by multinational corporations. More

importantly is that without activism many of these programs would not have been started,

so the more activism to keep corporations in check, the better.

Some examples in the apparel industry include Levi Strauss & Co. who set up

education and health services for their factories in Guatemala and their *Terms of

Engagement* initiatives in El Salvador.[lxxx] One of the most repulsive complaints from

laborers in South America is often about the factory cafeterias where food is often well

below health standards. Under cooperation with Levi Strauss, one of their factories

stopped serving food and instead set up programs with local vendors to supply food,

which extended community-building efforts in Guatemala. The adidas-Salomon

corporation has taken steps to initiate their *Standards of Engagement* programs in

Vietnam and Brazil.[lxxxi] In Brazil the company was particularly helpful in centralizing the dispersed home-based manufacturing into local factories so to better monitor problems such as child labor and safety issues. The company has also joined in efforts in Vietnam with Nike, Inc. and others, which are often downplayed by critics, but nonetheless have created worker enrichment programs.[lxxxii]

Visit just about any apparel company's Website and they will have a section on social responsibility, codes of conduct or labor issues. Does this mean we should run out and buy Nike or adidas shoes and Levi's jeans to reward their efforts? Most critics close to the issue would suggest opting for alternative purchases when possible. There is a list of shopping alternatives in the Consumer Solutions section of this book. Despite the sometimes-bold efforts of multinationals, full compliance with codes of conduct in their industry is something still at a distant reach. Corporations can legitimately publicize operations as socially responsible, but it is essential for activists and NGOs to stay on the supply chain trail. Especially so long as multinationals are free to seek out new export zones without accountability. For now it is just encouraging to be aware that these companies can be pressured in the right direction.

Big and Bad

Corporate watchdogs can maneuver a solid grassroots campaign and have reports issued about sweatshops in a matter of days. While it may take months to gather the data and reach workers within the factories willing to talk, word will travel fast when confirmed. This establishes an effective, yet limited, way to obtain some kind of transparency in secretive industries. The importance of transparency at the corporate level is also to ensure that corporations don't publicize whatever they want just to protect their

reputation. When Nike was bearing the brunt of much pressure put on by activists,

consumers and even government officials, this tactic backfired on them.

The company had set-up impressive programs in their large manufacturing operations

in Vietnam. It seemed like a genuine effort to respond to the pressure and put a

humanitarian investment into the regions they had reaped profits from. Nike often spared

no expense in publicizing these efforts and rightly so as the iconic swoosh image had

turned into the cultural swooshtika[lxxxiii] emblem of sweatshops. The problem came when

several NGOs including Global Exchange, Oxfam and CorpWatch, cracked the lies and

released counter reports to Nike's claims. The company retaliated yet again and ran

advertising in various media to counter the published NGO reports. Even with the ACLU

(American Civil Liberties Union) at its defense, Nike could not claim 1st Amendment

rights and the California Supreme Court found that Nike had indeed used misleading

commercial speech and advertising.[lxxxiv] While the content in question was editorial

letters in the *New York Times,* the NGO's had shown falsities in Nike's claims in

Vietnam. The company just repeated these false claims to the public in print, even after

being called out on them.

This is one reason why activists, NGOs, consumers and even congress resort to

babysitting efforts regarding social responsibility and corporations. The documentary

film *The Corporation,* establishes a scathing exposé of corporate misconduct with a

compelling analogy revealing we are not just dealing with a few bad apples.[lxxxv] While

we pick on Nike here as the bad apple of apparel, there are usually several labels sewn by

blood stained hands in the same factory.

It may seem somewhat extreme to resort to using imagery of bloodied hands, as activism campaigns often do. Though it is a reality workers face and one that organizations like the National Labor Committee have struggled to get across. Particularly with the factories producing goods for Disney in China, which furthers the frustration of when corporations say one thing and do another.[lxxxvi] The student activist group SACOM (Students and Scholars Against Corporate Misbehavior) in Hong Kong channeled their frustration mindfully by nominating Disney as the worst corporation of 2006.[lxxxvii] Such an award in academic and activist circles perpetuates Disney's reputation in corporate misconduct, leaving them little hope to reclaim it in a new economy. That is unless they actually live up to the codes of conduct they publicize and initiate programs that make it happen.

As critics of Disney, Nike, adidas, the Gap and other branded giants have pointed out; those codes look wonderful hanging on factory walls, translated into hundreds of languages, but are still ideals that must be reached. Also, without industry wide pressure and worker protection, displaying such codes may isolate workers. Case studies of factories producing for adidas-Salomon reveal, "...managers will make note of workers who talk to the inspectors or who even spend too much time looking at posted working condition standards..." preventing their presence from being truly effective.[lxxxviii] Progress is slow and it seems the frustration comes in knowing that multibillion-dollar operations are fully capable of raising the standards of pay a few cents to living wages. Or investing in factories or infrastructure that enables workers to live happier lives, which in turn translates into a motivated workforce.

Transparency and Cooperation

Corporate efforts to improve working conditions and also invest in community

building, when genuine, can create positive change. That is what is at the core of

improving conditions, bringing development to the community, not just the assembly

line. Still, individual companies cannot be held responsible for building social institution

programs, for reasons no one wants. Firstly it would put too much strain for, say a

branded shoe company, to take on the role of providing public services. Secondly,

workers do not want to feel their entire lives are controlled under the umbrella of one

corporation. This is already the case in some export processing zones where workers live

in compounds that produce for one label and literally eat, breath and sleep the brand. This

is the extreme of corporate influence in lives increasingly at the whim of globalization.

Corporations themselves have found that one label is not effective enough to pressure

even just one factory into improving standards. An executive from the Reebok

Corporation (before they merged with adidas-Salomon) had said, "We may be only 20

percent of a particular supplier's business and this has little influence, but we can

combine with two other companies that each have 20 percent, we can leverage our

influence over that factory operator."[lxxxix] While corporations may hold the threat of

pulling orders from factories, as globalization spreads so do the opportunities of

attracting other multinationals with cut-rate labor.

What needs to happen next is what the contributing authors of *Can We Put An End To

Sweatshops?*, unanimously agree on: total transparency. The authors' proposal of

"Ratcheting Labor Standards" (RLS) is essentially a way to pressure larger corporations

to reveal operations, which will in turn influence others to competitively, do the same.

The gateway to this domino effect is the principle of transparency, where if a free market

world is to exist fairly then it must be freedom that extends to information as well. The

authors Fung, O'Rourke and Sabel emphasize:

> *"The principles of transparency suggest a world in which consumers, workers,*
>
> *activists, and the public at large have the information they need to accurately and*
>
> *confidently identify initiatives to improve labor standards, gauge the results of those*
>
> *efforts, and compare the successes of firms, localities, and even nations against one*
>
> *another."*[xc]

This had an inkling of fruition in the start of monitoring programs like the Fair Labor

Association (FLA), Worker Rights Consortium (WRC) and The Clean Clothes Campaign

(CCC). Though these organizations are either limited in their influence, such as the CCC

or WRC, or influenced too much by corporate members, such as the FLA. In the case of

the FLA it was a step in the right direction, but hardly the independent monitoring of

factories that critics were demanding of the industry. For instance as a prominent FLA

member, Nike hired pricey auditing from firms like Ernst & Young and Price Waterhouse

who took necessary action, but critics largely deflated it as patchwork PR.[xci] In all

fairness, on any given day, Nike employs 550,000 people at over 22,000 factories in 52

countries[xcii], so if the company is genuine about enforcing manufacturing codes they

should welcome all the outside help they can get. Though independent monitoring

outside their control is something Nike and several other corporations resist.

These start-up efforts in monitoring were the result of an industry realization that

multinational corporations were not just buyers from distant markets. As one Nike

Executive said, "One of the biggest mistakes we made was to think we don't own the factories, so that's their problem. That's when we recognized we were more powerful

than we realized and, as a consequence, people expected more of us. Employees were

embarrassed and disenchanted and confused. The media has sweatshops and child labour

in every sentence."[xciii] When Gap, Inc made public its first social responsibility report in

2003, the Nike Corporation followed suit, an unforeseen effort to address sweatshops.[xciv]

While both companies, among others have been privately addressing the issue, this was

an unprecedented move in transparency. In the end corporations made up of people who

care about their impact on the world have nothing to lose in an initial stint of bad P.R. If

they act on the charges brought against them it will evolve into good press, as long as

reports of working with abusive factories lessen.

No matter the progress that corporations trickle out, as awareness of sweatshops

escalates concern, the reports of misconduct pile up. If major labels like the Gap, Nike,

adidas, and Wal-Mart cooperate with independent monitoring and act on criticism it can

prevent factories from lowering standards to stay competitive in pricing. This is the

pressure the apparel industry needs to impose instead of the pressure usually felt by

factories. As one Jordanian factory owner said about several multinationals, "they are

demanding us to be the lowest price, because they want to maintain everyday low prices,

we are not in business to be the lowest price, we are in business to sustain business."[xcv]

This pressure is part of what creates the fear that factory owners may have of independent

monitoring inspections, not the cultural relativism so often argued in the defense of

abuse.

Global Umpire vs. Empire

Independent monitoring is one way to ensure transparent compliance and perhaps the

most effective has been the SA8000 social auditing system of Social Accountability

International. In its trans-cooperative efforts across industries, SA8000 is connected to

monitoring the rights of over 15 million workers.[xcvi] If such a system could be harnessed

to provide assurance to consumers at the point of sale, the market of fair trade goods is no

longer limited to awareness and extends to standardization. Every program is subject to

critics, as Deborah Leipzinger weighs it, "Criticism of SA8000 varies and sometimes

appears contradictory. For Example some criticize SA8000 for being very rigorous,

whereas other regard is as being too lenient. For some, it is a strength that SA8000

requires investment, whereas others regard SA8000 as 'expensive.'"[xcvii] The price

incurred by the manufacturing plant requires a 3-year commitment where audits can cost

from $3,000 to $5,000 including all the expenses.[xcviii] The investment is in that factories

will be given preference if SA8000 certified, or may even be required to do so. If the

factory can comply within the 3-year time frame, it will position them to develop long-

term working relationships with multinational corporations under pressure to end

sweatshops.

This structured form of monitoring can free corporations from tasks they did not want

in the first place. It is fair to say that should a corporation not pursue or agree to

independent monitoring then they are hiding something. A seemingly simplistic claim,

but aside from initial cost, what does industry have to lose in building a platform of

international labor standards. Though this does not free corporations entirely from the

responsibility of embracing factories as an integral link to be valued above profits. Part of
the radical changes that activists want to see is for corporations to reflect something
beyond the profit potential and expansion in globalization. The far greater realization
comes in knowing that capital is a bridge to building relationships around the world that
replaces tension with trade. When distant partners go into business together the
xenophobia that has infected the last century loses out to social progress. The last thing
western multinational corporations want is to perpetuate a reputation as just that; a
western multinational corporation that likes to suppress labor costs with sweatshops.

Leaders of multinationals have reached public acknowledgment of their influence over
governments and world bodies; some even accepted responsibility as global citizens. This
will have the greatest impact in a country such as the United States, in what C. Derber
calls Umpire Economics. He describes the U.S. influence over world trade as an umpire
at a baseball game, but one that bends the rules to suit certain teams.[xcix] Those teams are
U.S. multinationals that use the weight of profits to tip the scales of trade advantage. This
adds to the international perception of the American Empire, which since World War II
has gained much scholarly evidence as imperialism through economics. If corporations
involved themselves in trade reform, instead of manipulating it behind democracy
wrapped in barbed wired fence, it can give world bodies their rightful role as umpire.

This can build the true foundations to the bridges that will ease national tensions in
trade and eventually over the dilemma we will face in natural resources. The system of
trade is the bridge that exists now, but both sides need the right of away to freely cross it.
At its core this element of the economy requires understanding not just the needs of
workers, but in how cultural differences are often exaggerated to harmful ends.

The Problem of Standards

Following the benefits of transparency does more than just keep activists at bay or appease consumers. Recognizing the greater purpose, Jill Murray observes, "…that the system of international labor law does not provide a set of clearly defined rules that firms can apply to their individual workplaces in any jurisdiction in the world." This is the problem of standards, but is overcome by the process of awareness through transparency, as Murray adds "…what is learned by one firm can be spread among others in similar positions."[c]

Awareness is not only needed from Western firms in understanding the basic human needs of foreign workers, but in encouraging factory owners to bridge cultural gaps. This is a problem of relativism that can create unrealized tensions in the workplace between owners and laborers. For instance many times owners can often be, say a Korea national that runs a factory in Thailand or even as far as Mexico. If the global supply chain is to continue in transnational expansion, care must be taken to understand customs that may in fact find more commonality than differences between firms, owners and workers.

The most available organism in giving workers their voice is unionization, which faced from atop the corporate ladder has a bad rep. A non-unionized factory is almost certainly a sign of a sweatshop in a region of low development. It is often believed that unions are a fierce organism that must demand rights from oppression and that is often what they arise from. There is still a more interlinked aspect of unions in that they can organize workers into a position that creates more ease of communication to the owners and distant corporations. Guy Standing of the ILO places unions in precedence over the advancement of social auditing programs like SA8000. Through public advocacy, he

states, "strengthening the voice of working communities…is the most effective way to

make substantial progress."[ci] From the roots of community workers can in fact establish

their own standards of rights and develop incentives for both their productivity and

compliance by the factory. This can preempt what Standing fears is developing a need for

factories and multinationals to comply with complex standards and monitoring. The

typical system of punishment and quotas is a form of oppression that creates a

psychologically damaging master-slave relationship. Unions are empowerment from this

state and multinationals can create advocacy programs that encourage, or pressure,

factories to allow unions. This will save them the cost of monitoring and also the industry

having to force standards that can't apply universally. Again, here is something that

demands great leaps of foresight and flies in the face of traditional corporate suppression

of labor.

Allowing Complexity to Avoid It

Allowing the kind of development from the bottom up makes the job of applying

standards down the road much easier. At that point a system like SA8000 can have its

strongest impact, where regions comply within their own standards of operation to meet

international expectations of labor rights and product quality. This appears complex from

the start, but can be an organic process that lets the details of reality fall into place of the

elevated ideals. Though unions may hold a more pressing importance in setting standards,

it may be easier for them to maintain standards with a monitoring system in place.

There is also an ethical dimension in growing the corporate workforce that companies must be aware of now. It extends to the manufacturing arm of operations from the employees that make up the corporate activity of firms. The old notion, good help is hard to find, has taken on new meaning and human resource departments must ensure an organization's social responsibility to attract talent. Hartman, Arnold and Waddock agree that, "some evidence suggests that potential employees increasingly care about the ethical reputations of the companies for whom they work, such that it costs more for low-reputation companies to hire high-quality employees."[cii] This is just another detail that can fall into place if corporations open their awareness to having an impact by just listening to the tremors of change ahead.

This may be a change towards more small business operations, as prospective employees who can't find corporations they believe in just start their own. What this means for larger multinationals is less control and rapid decline of competitive advantage. No matter how the big the legs of the giant, he can't stamp out all the seedlings of business in a do-it-yourself culture. Just the advent of independent media has grown to defuse the smoke and mirrors of corporately controlled news and entertainment. This in itself is what often allows reports of sweatshop abuses to break through the often one sided media ownership of globalization's imperialists. What such phenomenon create is decentralized social organisms, seemingly complex in their abundance, but giving us better understanding of the world.

◙◙◙

[lxxvi]Charles Derber notes in his book *People Before Profit* (Picador, 2003) that "the Fortune 500 global corporations, while hogging about one-quarter of the world's income, employ less than 1 percent of the world's workers." (pg. 178).

[lxxvii] Norberg shows in his book *In Defense of Global Capitalism* (Cato Institute, 2003), "Between 1980 and 1993, the 500 biggest American firms saw their share of the country's total employment diminish from 16 to 11.3 percent," and "the 50 largest corporations in the world have a GDP equaling only 4.5 percent of that of the 50 biggest countries." (pg. 215)

[lxxviii] Charles Derber, *People Before Profit*, (Picador, 2003), pg. 60

[lxxix] Charles Derber, *People Before Profit*, (Picador, 2003)

[lxxx] Tara J. Radin, "Levi Strauss & Co.: Implementation of Global Sourcing and Operating Guidelines in Latin America" in *Rising Above Sweatshops*, ed. Hartman, Arnold and Waddock (Praeger, 2003), pgs. 250 – 282

[lxxxi] Hartman, Wokutch and French, "adidas-Salomon: Child Labor and Health and Safety Initiatives in Vietnam and Brazil" in *Rising Above Sweatshops*, ed. Hartman, Arnold and Waddock (Praeger, 2003), pgs. 191-247

[lxxxii] Hartman and Wokutch, "Nike, Inc.: Corporate Social Responsibility and Workplace Standard Initiatives in Vietnam" in *Rising Above Sweatshops*, ed. Hartman, Arnold and Waddock (Praeger, 2003), pgs. 146-189

[lxxxiii] The Swooshtika is a activist logo portraying four Nike swoosh logos that form the shape of a swastika.

[lxxxiv] "The California Supreme Court Rules Nike Can Be Sued for Lying About Its Labor Practices, But the AcLu Sides with Nike: A Debate" Narr. Amy Goodman, *Democracy Now*, New York, May 8th, 2002, http://www.democracynow.org/article.pl?sid=03/04/07/0258215&mode=thread&tid=5 (Accessed, 4/06)

[lxxxv] *The Corporation*, Dir. Mark Achbar, Jennifer Abbott and Joel Bakan, Big Picture Media Corporation, 2003

[lxxxvi] The National Labor Committee, "Disney's Children's Books Made with Blood Sweat and Tears of Young Workers in China", http://www.nlcnet.org/news/china_info.asp (Accessed, 8/06)

[lxxxvii] Farzin Mojtabai and Jason Cangialosi, "Disney Corporation Given 2006 Public Eye Award", Mar. 6th 2006, http://www.associatedcontent.com/article/23902/disney_corporation_given_2006_public.html (Accessed, 8/06)

[lxxxviii] Hartman, Wokutch and French, "adidas-Salomon: Child Labor and Health and Safety Initiatives in Vietnam and Brazil" in *Rising Above Sweatshops*, ed. Hartman, Arnold and Waddock (Praeger, 2003), pg. 211

[lxxxix] Michael A. Santoro, "Philosophy Applied I: How Nongovernmental Organizations and Multinational Enterprises Can Work Together to Protect Global Labor Rights" in *Rising Above Sweatshops*, ed. Hartman, Arnold and Waddock (Praeger, 2003), pg. 114

[xc] Fung, O'Rourke and Sabel, "Realizing Labor Standards" in *Can We Put An End To Sweatshops*, (Beacon Press 2001), pg. 19

[xci] Dara O'Rourke, "Vietnam: Smoke From a Hired Gun, A Critique of Nike's Labor and Environmental Auditing in Vietnam as performed by Ernst & Young", Nov. 10th 1997, CorpWatch, http://www.corpwatch.org/article.php?id=966 (Accessed, 5/06)

[xcii] Hartman and Wokutch, "Nike, Inc: Corporate Social Responsibility and Workplace Standard Initiatives in Vietnam" in *Rising Above Sweatshops* ed. Hartman, Arnold and Waddock (Praeger Publishers, 2003) pg, 145

[xciii] Nicholas Ind, "A brand of enlightenment", in *Beyond Branding: How the New Values of Transparency and Integrity Are Changing the World of Brands*, (Kogan Page, 2003), pg. 10

[xciv] Farzin Mojtabai and Jason Cangialosi, "Gap, Inc. and Sweatshops", June 2nd 2006, http://www.associatedcontent.com/article/33681/gap_inc_and_sweatshops.html

[xcv] Farzin Mojtabai and Jason Cangialosi, "U.S. Senator Byron Dorgan's Anti-Sweatshop Bill", June 5th 2005, *http://www.associatedcontent.com/article/41123/us_senator_byron_dorgans_antisweathop.html?page=3*

[xcvi] The Social Accountability International Website states, "SAI works with companies, consumer groups, non-governmental organizations, labor organizations (which currently include a total of 15 million workers in their ranks), governmental agencies, and certification bodies around the world. SAI accredits the certification bodies for SA8000 auditing to ensure that workers receive the just and humane treatment they deserve." http://www.sa-intl.org/index.cfm?fuseaction=Page.viewPage&pageId=472 (accessed 8/06)

[xcvii] Deborah Leipzinger, *The Corporate Responsibility Code Book*, (Greenleaf Publishing, 2003), pg. 157

[xcviii] S. Prakash Sethi, *Setting Global Standards: Guidelines for Creating Codes of Conduct in Multinational Corporations*, (John Wiley & Sons., 2003), pg. 229

[xcix] Charles Derber, *People Before Profit*, (Picador, 2003), pgs. 80-104

[c] Jill Murray, "The Global Context: Multinational Enterprises, Labor Standards, and Regulation", in *Rising Above Sweatshops*, ed. Hartman, Arnold and Waddock (Praeger, 2003), pgs. 41 and 42

[ci] Guy Standing, "Human Development" in *Can We Put An End To Sweatshops*, (Beacon Press 2001), pg. 79

[cii] Hartman, Arnold and Waddock, "Rising Above Sweatshops: An Introduction to the text and to the Issues" in *Rising Above Sweatshops*, ed. Hartman, Arnold and Waddock (Praeger, 2003), pg. 6

Chapter 6

Consumer Solutions:

Purchasing Power

When dealing with any issue facing the world today it often begins at home. Ironically this may be the danger zone many face in tackling the issues that take big efforts to wrap your head around. Plenty of student activists confront the dilemma of returning home an angry college student and social activists come off as just crazy mad at the world. Though as this book has built an argument from a historical context and from within the current state of economic affairs, there is a target to toss the dart of blame towards. This can leave us, as consumers, complacent that there is no way to affect change on big problems. Though to really fire off the arrows of change the BullsEye starts with personal decisions and then works toward collective ones.

Thinking about the massive global numbers in terms of purchasing power and markets that dictate policy, can make the potential for change evident. We can all march off to sprawling protests and say we boycott certain brands or buy locally, but this is only the beginning. Thankfully grassroots activism has created accessible ways for students, thinkers and the public at large to engage more than individual consumer solutions. The relationship of civic duty to humanity and being a responsible consumer, while not something your friends want to be preached to about, will join you when they see the change. This requires you, the individual, to walk, talk and be the changes that burn inside your mind.

That of course comes from perhaps the most powerful labor movement in recent history as Gandhi walked with his people to create change. This happened as the civil rights movements grew in the United States and the feminist growl turned into a roar. It also chimed into what would culminate into the war protests, as the military-industrial complex made its way into Vietnam. All of these movements cannot be overstated as influences in the grassroots democracy that is returning to the public mind today. The Anti-sweatshop movement beams from within this desire for change, with a bold, accessible agenda. Today it is as much a part of the consumer culture as the brands it hopes to raise the issue to. Though as noted throughout the book the only language economists and corporations listen to comes from the hard earned money we part with. This is what sets consumers in a pivotal role in an already loud debate and what we buy or don't buy turns up the volume.

As Linda Golodner, President of the National Consumers League reflects back "Sweatshops and Child Labor are not new concerns nor a new battle for consumers."[ciii] She is referring to the "White Label" movement of America in the early 20-century, even early as 1891, where consumer activists were in solidarity with garment workers. They adorned clothing with a white tag to signify that the garments had been made in factories that complied with workers' demands. Sweatshops infested the eastern cities of the U.S. at the heights of industrialization and continued on as factories sought states where they could operate free of unions. This early movement established an American consumer ethic that linked the world of consumption and production. This ethic was lost into the outsourcing waves of labor, disconnected by oceans of distance and trade policy agenda.

What this historic gesture of solidarity shows is that people are always willing to rise above their roles as consumers to give lift to the role of labor. When union strikes happen here in the U.S., whether factory workers, teachers or public transit, cities feel the impact. If unions were denied jobs protection, chances are others would join in the struggle even just to protect the principle of fair labor. That the plight of global workers is largely unheard or ignored here in the U.S. has many factors against it, but largely it is consumers disconnected from global productions. Trade keeps goods flowing into the country and with labor practices out of sight (out of mind), the cheap prices keep people content and buying.

As mentioned sweatshop activism often gets criticized as a conspiracy of labor unions in the U.S. trying to keep domestic jobs and maintain standards lost in outsourcing. Consumers can readily relate to the issue of job loss in outsourcing, as it might be there job next. That claim is being answered by the issues of education reform and new job creation in the U.S. Meaning that while manufacturing jobs have been lost, the greater importance of higher education standards needs to meet the demand of an advancing global economy. This has led many talking head experts to say that the U.S. is at risk to losing its superpower status to China, India and the European Union. Upon closer inspection in the context of globalization, this can really imply that power is beginning to realize increased distribution. While still far from equalizing trade, capital and job creation, it may yet be distributed in a way that balances the global economy from its centralized system.

The culture of consumerism also spreads with this power distribution, something

frequently noted in Japan, China and India. This comes from rising middle classes that

allow shopping centers to grow around metropolitan areas with job growth. Positive

aspects of this growing middle class stimulate the discussion that free trade in

globalization brings prosperity over time. As the economies tout a massive population

boom and corporations maneuver trade agreements into freshly tapped markets, the

effects of consumption at current rates will take a toll. A toll on community as people are

reduced to consumers, a toll on the environment as more resources are drained and a toll

of the labor regions that must be suppressed to maintain affordability in consumer

demand. The world is at a transitional phase and most of the alarm is over the unequal

growth in trade. The growing awareness of such large issues sets people into a stance of

preparation for change. In this sense consumers are ready to buy into something new.

Faith

Strangely, something that liberal economists and free market advocates share with

their critics who emphasize fair trade philosophy is faith in the market. Scholar Raul C.

Panglangan touches upon this in describing the approach in sweatshop activism that

"bypasses state-based mechanisms altogether in favor of the market, operating, for

example, through social labeling, voluntary corporate codes of conduct and NGO-led

boycotts of tainted goods, thus enabling consumers to vote their consciences with their

pocketbooks."[civ] This resembles the corporate culture that encourages purchasing power.

One that has fought to escape the clutches of governments so their created mechanisms allow them to run on their own laws. Consequently activists have fought to escape the clutches of corporate dominance and government inaction.

Simply waiting for corporations to initiate a system of fair labor within their operations does nothing. As consumers, we are aware that certain firms have been exposed as using sweatshops. It cannot be taken on faith that just raising the issue will create change for the workers and pressure must come from many angles. This is especially important if we continue to buy the products that support the continuing use of sweatshops. With the evidence stacked before us, it is not just guilt that prevents us from supporting a certain company; it's a demand for change. More importantly, if our purchasing decisions are brushed aside as only guilt it is easily overridden by the seduction of want in a world of plenty.

As mentioned in this book, The Marymount University survey of consumers, apparel and sweatshops is quoted in just about every text on the topic published in the past five years. This testifies the impact of the survey as reassurance that consumers would support corporations that made efforts in improving conditions. It also shows the "moonwalking" factor mentioned in chapter 4, as there is no real evidence as far as apparel to show numbers of ethical purchasing power. Instead social institutions gather comparisons and surveys that will hopefully convince the corporate world of the potential. John Dunning relishes that "Consumer activism is, indeed, very much alive", citing a Gallup poll in 1995, "that found three out of five UK consumers were prepared to boycott stores or products because they were concerned with the ethical standards of the suppliers." Another UK poll "showed that three quarters of respondents made their choice of products on a green or ethical basis."[cv]

Is this to say that people have become more ethical over the years and are just now looking for those options as consumers? Or it may be that the world has awakened to the potential of change with their dollars. This follows from the rising middle classes around the world who now feel empowered by their purchases. People may want their purchases to mean something more than just consumption. The realization of this depends on the ability to invest in the things that are meaningful, like community, which prompts many to buy locally. Without community, C. Derber warns, "democracy is out the window, and the possibility of the good life rapidly erodes, as people try to compensate for the loss of human connection with the thrills of headlong consumerism."[cvi]

The evidence is enough for corporations to listen in for the potential of a market, but they have not moved on the market. This is partly because it is in the best interest of consumers, workers and activists to ensure that we are not duped. For corporations to truly affect change in labor standards it requires greater strides than they have taken. We could get duped in that a simple label could proclaim fairness in efforts from corporations and not achievements. Also keep in mind that corporations have genuine reasons for hesitation at a business level outside of moral imperatives. Amy Domini mentions how, "Even the most humane purchasing agent sitting in her New York City office is unlikely to notice the human rights implications of what appears to be a perfectly straightforward economic decision."[cvii] The advent of fair trade in apparel as a market is a tremendous leap from a consumer apparel trend that believes the cheaper the better. One scholar notes, "Despite the fact that opinion polls have probably over-estimated the numbers of consumers who will buy from the ethical apparel manufacturers and retailer, it is likely that at least some consumers shop in altruistic versus self-interested ways."[cviii] As hopeful as this may seem, the operative word is "some" and not all. This explains a troubling

problem in purchasing power, revealing a market that depends on morals and not just

demand.

Here again the underlying mechanisms of economies are exposed not as the natural

laws they are so thought to be, but as belief or faith in markets. In many respects the

consumers who have come to question the goods they are sold, have become

disenfranchised to the point of conservation, or learning to live without. So regardless of

growth areas and population rises, corporations lose consumers who don't believe

conditions can improve for workers and corporations lose faith in consumers to support

such a change.

The Cost of Change

Looking at the potential for such a market isolated from the structure of government

or social accountability lessens its prospect as well. Every party involved in the supply

chain will incur some portion of the costs involved initially. This essentially is an

investment and at some levels an operating expense. Everyone involved must want the

same thing, and going back to Adam Smith we remember that moral capitalism is the

only true free trade. Workers want better conditions, whether they raise their voices to it

or not, we can all relate to realizing entitlement to certain basic rights. Consumers want

the workers to have these improved conditions, as they cannot justify a selfish "us and

them" mentality. This is the moral bomb often dropped on arguments about sweatshops;

just ask anyone if they would want to work in such conditions. Corporations want to

succeed by providing consumers what they want and they cannot "operate successfully

without the tacit consent of its customers, the local community and society at large."[cix]
This includes government and if a true democracy is to prevail it must reflect these needs.
It is best to focus intention on what we want to exist here before we get buried under the
truth of what is. This defeats us into submitting to purchases we may well know supports
something we don't believe in as moral, rational people.

When we talk about cost here, it is not in unreasonable investments towards
improvement. In most cases it is mere thousands of dollars in the multibillion-dollar
operations of multinationals. This has seen widespread agreement, such that, "For the
manufacturers of brand-name retail goods, a significant increase in labor costs may be
readily absorbed as an operating expense. Indeed, the expense may be offset by the value
added to the good insofar as the consumers demonstrate a preference for products
produced under conditions in which the rights of workers are respected."[cx] Theodore
Moran concurs in his book *Beyond Sweatshops* that, "It is not unreasonable to
conclude…that most, if not all, of the extra expenses associated with better treatment of
workers could be absorbed by investors, retailers, and consumers in the developed world
without producing a large reduction in employment (or substitution of capital for labor)
in developing countries."[cxi] From a logistical standpoint the costs of ending sweatshops,
with some refinement, is not an overwhelming feet that corporations will struggle with.

Consumers may have to fork over an extra dollar, but in reality it may be even less
with price reductions as the market grows. Comparatively, prices of goods made under a
fair trade label, whether shoes, jewelry or coffee is competitive when available.
Accessibility is the largest dilemma in really promoting a program for buying fair trade

goods, as well as limited selections. This is the initial hurdle many see in getting the

market to react to having a choice and letting the surveys become reality.

Demand vs. Deception

One question these surveys can lead us to is do we really have a choice? For many

people, it is a red state or blue state, Nike or Reebok, Pepsi or Coke, revealing that

demands we supposedly have are created for us. As they say in politics, we resort to

choosing between the lesser of two evils and in many ways even our vote has really

become nothing more than another idea marketed to us. This false message fed to the

American public extends from the voter booth to the shopping mall. It is necessary to

look beyond this common belief, expressed one way by Johan Norberg that, "Nothing

forces people to accept new products. If they gain market share it is because people want

them."[cxii] There is plenty of reason to go beyond this economic law and look at the

cultural impact of advertising and media. Our public space is clustered with these driving

forces of demand, but many have begun to observe it as deception.[cxiii] Part of the anti-

sweatshop movement addresses this element of consumerism and calls on people to be

more than the brands they buy.

It is often hard for people to place themselves within such a position without feeling

anti-capitalist or anti-corporate. Both arguments have valid claims, but as globalization

expands it's a bit like saying you're a fish without gills while swimming in the sea. Liza

Featherstone also asks, "…one wonders whether anti-corporatism is really about social

justice, or simply an aesthetic objection to bigness."[cxiv] This type of activism reaches

across the U.S and the world as communities reject the big box chain stores from invading their towns. The entire country of India has a ban against such stores, even

though they are one of the largest exporters of goods to these very stores. Though in 2005

Wal-Mart began plans for moving into the country, which may say something about

India's changing face of consumerism.[cxv] There is something deeply amiss about that and

whether social justice or aesthetic preference, the arguments find intersects that can

compliment one another.

By acknowledging that the economic demands we support are not entirely, and very

often far removed from the necessities of life, we join this aesthetic. From there we see

that the connections to social and environmental issues are tangled at the roots beneath

the surface and in need of constant cultivation. This comes at a personal level first as we

cultivate our minds to realize the kind of world we want to live in. Then our choices, no

matter how small, become apparent in having an affect on something like the global

economy. The final picture created from this realization can show that the global gap of

poverty does indeed exist and we perpetuate it by pursuing the deceptions that misled us

as demands.

The Road Ahead

One way to avoid angry activist syndrome is to use a meditative technique of self-

observation before action. Not all activism is proactive in creating change and can

become redundant and even counterproductive. For the most part anti-sweatshop activists

have honed in on worker solidarity and pushing for higher labor standards. This

encourages workers to stand their own ground and provides the most effective monitoring

any corporation could have, its workers. Over time the challenge activists present to corporations can become an allied effort. Activists must see themselves as consumers and

consumer must see themselves as activists. This may offer a better way of seeing a

seemingly deceptive corporate culture as more of the world for what it is. Yes, it is full of

greed and the capitalist system gives almost free reign to purse it. The friction we feel in

acting morally within that system is activism in motion. The motion that is felt the most

is what comes from consumption, as the system is an economic one. If activists detach

themselves from the corporations they challenge, they no longer hold the threat of lost

consumption. This comes back to the difficulty of proving that consumers will use their

label conscious instincts and choose goods that make ripples of change. The world of

activism and consumerism must meld into a demand for fairly traded goods.

The concerns over creating a complex labeling system will be greatly avoided if

consumers push for systemic change. Analysis shows, "a more generic system might

have an even stronger impact, if major retailers in, say, the garment, footwear, toy,

jewelry, and hardware businesses—including large discount outlets—chose to offer only

sweatshop-free products, it would become genuinely inconvenient for even unconcerned

consumers to buy goods produced in substandard or un-inspected plants."[cxvi] The

initiative of consumers to voice this to the corporations and stores, whether as protest,

letters and even verbal communication on the sales floor is needed, but hardly enough.

We have the vehicles of change available to us through governments, but the fears of

consumers and corporations in government control must be overcome. Realizing that it is

often corporations that influence bad political policy and activity, we have the power to

make the governments our own. This comes not only in voting, but also in expressing

that the real demand we have is for a better world through fair labor and trade.

If social justice movements can influence the economy of the second most widely

traded commodity on the planet, coffee, then it can be true of almost any market. The

labeling system that is developing from the fair trade coffee market speaks volumes, but

is not foolproof. The seduction of falsely labeling a product as ethical to charge more for

it exists in a system competitively seeking those prices. People within corporations must

also view themselves as the global labor force and ask what they would want out of the

economy. This rather simple ethic can lead corporations to work with an independent

labeling system that ensures fairness. In time, in conjunction with the advent of Socially

Responsible Investing (SRI), efforts can organize the global economy into a system we

can not only live with, but also be proud of.

Building up locally and supporting independent producers is a form of activism in

itself, but not the solution to sweatshops. It attempts to overstep the cause, but it is a leap

too large to make in that single effort. Opening a share of the market for independent

producers enhances the economy, but our existing system needs to be fixed along with it.

A straight boycott of brands may at first raise awareness, but it does not help the problem

of development in countries that need jobs. Several of the leading organizations in the

Anti-sweatshop movement, namely the National Labor Committee (NLC) and the Clean

Clothes Campaign (CCC), don't support boycotts. Andrew Ross cautions that,

"...boycotts can punish the most vulnerable workers, when plants shut down in

communities that need jobs and wages."[cxvii] At the same time Ross has also admitted that,

"Banking on consumer conscience had paid off in a number of campaigns. For instance,

in May 1995, the Child Labor Coalition mobilized a consumer boycott of Bangladeshi

clothing exports after investigations revealed the widespread use of child labor in the industry."[cxviii] This was a national boycott that got governments involved and the pressure

was too great for Bangladesh to just sit it out and wait.

The more effective route for a widely networked organization like the CCC is to run

"focused consumer campaigns on the 'Labour behind the label' whose goal is mobilizing

the purchasing power of particularly young consumers," notes Michele Micheletti.[cxix]

Firing up the young to get fair in their fashion has a certain appeal in that people are

inspired to wear their beliefs. Though the change that will command the global supply

chains into this fashionable harmony is formed in procurement. Marieke Eyskoot of the

Clean Clothes Campaign wrote that "Governments don't just make laws; they are also

consumers," in her 2006 Sweatfree Communities International Conference

presentation.[cxx] The public procurement of goods for both national and local governments

is one of the largest apparel markets and one largely overlooked in the brand wars.

The steam behind several U.S state laws sought to ban the purchase of government

apparel from sweatshops. This goes back to the child labor advocacy of Senator Tom

Harkin (D) of Iowa who introduced several congressional efforts to prevent sweatshops.

One of which prompted President Clinton in 1998 to issue an executive order to stop the

Federal government from using goods made by forced and indentured labor.[cxxi] This also

extends to 2006 where the human trafficking in Jordanian sweatshops moved Senator

Byron Dorgan to issue *The Decent Working Conditions and Fair Competition Act.* Such

legislative action pressures foreign governments to enforce labor standards or lose out to

valuable western markets. While some may claim that such efforts impose tariffs

detrimental to free trade, it heightens awareness of the issue into action. It also exercises

the democratic process, which would have a far greater impact were more to use it.

Corporations have no choice but to comply with the demands of consumers and workers in solidarity with democracy. Initially, "The presence of multinational corporations in oppressive governments can very often be an aid to the pursuit of democracy, because those corporations are sensitive to pressure from Western consumers, which has a direct impact on sales," as Norberg argues.[cxxii] This can only be effective if corporations are under legal pressure to comply and not relying entirely on unproven purchasing power. The process depends on that corporations are firstly accountable, who can then initiate the improvement of a country's labor standards. Furthermore we cannot rely upon the notion that purchasing power will spread democracy through globalization. Derber cautions that, "As a sovereign principle, one dollar, one vote, is inherently undemocratic, and it ensures a growing gap between the rich and poor because it gives the rich far more political representation."[cxxiii] The all mighty dollar may tell corporations what consumers prefer and influence policies of trade and labor, but it is not an individual's voice.

In the end we take our roles as consumers only so far and we cannot escape the problems of the world in what we buy. Not only in escapism through consumerism, but also in putting all our faith in the label. That is what troubled activists in the first place and our trust in the friendly, often inspiring labels of the products betrayed us. For those that want more equality and fairness in the world, realizing that the systems of trade provide the greatest vehicle begins at home. To rise above the disturbing illusion that we are what we buy, we must create an active role beyond consumption. If that empowerment is connected to globalization, what and how we trade will be the life force of a peaceful co-existence.

◎◎◎◎◎

[ciii] Linda Golodner, "Consumer Solutions and Action", in *Child Labor and Sweatshops*, ed. Mary E. Williams, (Greenhaven Press, 1999) pg. 57.

[civ] Raul C. Pangalangan, "Sweatshops and International Labor Standards", in *Globalization and Human Rights*, ed. Alison Brysk, (University of California Press, 2002) pg. 98.

[cv] John H. Dunning, "The Moral Imperatives of Global Capitalism: An Overview", in *Making Globalization Good: The Moral Challenges of Global Capitalism*, ed. John H. Dunning, (Oxford University Press, 2003), pg. 28

[cvi] Charles Derber, *People Before Profit*, (Picador, 2003) pg. 134

[cvii] Amy Domini, *Socially Responsible Investing: Making a Difference and Making Money*, (Dearborn Trade, 2001), pg. 44

[cviii] Marsha A. Dickson, "Identifying and Profiling Apparel Label Users", in *The Ethical Consumer*, ed. Rob Harrison, Deirdre Shaw, Terry Newholm, (Sage Publications, 2005), pg.158

[cix] Hartman, Wokutch and French, "adidas-Salomon: Child Labor and Health and Safety Initiatives in Vietnam and Brazil", in *Rising Above Sweatshops*, ed. Hartman, Arnold and Wokutch (Praeger, 2003), pg. 207

[cx] Denis G. Arnold, "Philosophical Foundations: Moral Reasoning, Human Rights, and Global Labor Practices" in *Rising Above Sweatshops* ed. Hartman, Arnold and Wokutch (Praeger, 2003), pg. 93

[cxi] Theodore H. Moran, *Beyond Sweatshops: Foreign Direct Investment and Globalization in Developing Countries*, (Brookings Institute Press, 2002), pg. 57

[cxii] Johan Norberg, *In Defense of Global Capitalism*, (Cato Institute, 2003) pg. 213

[cxiii] For cultural and economic studies of marketing and advertising see Naomi Klein's book *No Logo* (Picador, 1999) and Douglas Rushkoff's book *Coercion* (Riverhead, 1999). Also see the documentary *The Corporation*, which interviews Klein. Rushkoff's Frontline PBS documentary, *The Merchants of Cool*, also interviews Klein.

[cxiv] Liza Featherstone, *Students Against Sweatshops*, (Verso, 2002), pg. 34

[cxv] Forbes notes that India exports over $1.5 Billion in good to Wal-Mart, and purchases $0.00 from the store. http://foreignexchange.tv/?q=node/1508 (Accessed 8/06), also Robert Malone, "Wal-Mart Reshapes the Retail World", *Forbes*, 1/24/2006, http://moneycentral.msn.com/content/invest/forbes/P142021.asp (Accessed 8/06)

[cxvi] Theodore H. Moran, *Beyond Sweatshops: Foreign Direct Investment and Globalization in Developing Countries*, (Brookings Institute Press, 2002), pg. 96

[cxvii] Andrew Ross, *Low Pay High Profile: The Global Push for Fair Labor*, (The New Press, 2002), pg. 48.

[cxviii] Andrew Ross, *No Sweat: Fashion, Free Trade and the Rights of Workers*, (Verso, 1997), pg. 41

[cxix] Michele Micheletti, *Political Virtue and Shopping: Individuals, Consumerism, and Collective Action*, (Palgrave Macmillan, 2003), pg. 98

[cxx] Marieke Eyskoot, "Clean Clothes Campaigning on Ethical Public Procurement in Europe", in *Report: Sweatfree Communities International Conference 2006*, pg. 9, http://www.sweatfree.org/docs/conferencereport2006 (Accessed 8/06)

[cxxi] "Child Labor" and Senator Tom Harkin, http://harkin.senate.gov/child-labor/index.cfm (Accessed 7/06)

[cxxii] Johan Norberg, *In Defense of Global Capitalism*, (Cato Institute, 2003) pg. 222

[cxxiii] Charles Derber, *People Before Profit*, (Picador, 2003), pg. 78

Index

www.ingramcontent.com/pod-product-compliance
Lightning Source LLC
Chambersburg PA
CBHW022123280326
41933CB00007B/525